**Milestones in Drug Therapy
MDT**

Series Editors
Prof. Michael J. Parnham, PhD
Senior Scientific Advisor
PLIVA Research Institute Ltd
Prilaz baruna Filipovića 29
HR-10000 Zagreb
Croatia

Prof. Dr. J. Bruinvels
Sweelincklaan 75
NL-3723 JC Bilthoven
The Netherlands

Sildenafil

Edited by U. Dunzendorfer

Springer Basel AG

Editor
Priv.-Doz. Udo Dunzendorfer
J.-W.-G.-University
Zeil 65–69
60313 Frankfurt/Main
Germany

Advisory Board
J.C. Buckingham (Imperial College School of Medicine, London, UK)
R.J. Flower (The William Harvey Research Institute, London, UK)
G. Lambrecht (J.W. Goethe Universität, Frankfurt, Germany)

Library of Congress Cataloging-in-Publication Data
Sildenafil / edited by U. Dunzendorfer.
    p. ; cm.-- (Milestones in drug therapy)
Includes bibliographical references and index.
ISBN 978-3-7643-6255-3   ISBN 978-3-0348-7945-3 (eBook)
DOI 10.1007/978-3-0348-7945-3

1. Sildenafil. 2. Impotence--Chemotherapy. I. Dunzendorfer, U. (Udo), 1944- II. Series.
[DNLM: 1. Impotence--drug therapy. 2. Piperazines--therapeutic use. 3. Cardiovascular Diseases--drug therapy. WJ 709 S582 2004]
RC889.S525 2004
616.6'922061--dc22

2004052996

Bibliographic information published by Die Deutsche Bibliothek
Die Deutsche Bibliothek lists this publication in the Deutsche Nationalbibliografie; detailed bibliographic data is available in the Internet at <http://dnb.ddb.de>.

ISBN 978-3-7643-6255-3

The publisher and editor can give no guarantee for the information on drug dosage and administration contained in this publication. The respective user must check its accuracy by consulting other sources of reference in each individual case.
The use of registered names, trademarks etc. in this publication, even if not identified as such, does not imply that they are exempt from the relevant protective laws and regulations or free for general use.
This work is subject to copyright. All rights are reserved, whether the whole or part of the material is concerned, specifically the rights of translation, reprinting, re-use of illustrations, recitation, broadcasting, reproduction on microfilms or in other ways, and storage in data banks. For any kind of use, permission of the copyright owner must be obtained.

© 2004 Springer Basel AG
Originally published by Birkhäuser Verlag in 2004

Printed on acid-free paper produced from chlorine-free pulp. TCF ∞
Cover illustration: Molecular structure of sildenafil (see p. 18).

ISBN 978-3-7643-6255-3

9 8 7 6 5 4 3 2 1   www.birkhauser-science.com

Contents

List of contributors	VII
Preface	IX

Ian H. Osterloh
The discovery and development of Viagra® (sildenafil citrate) 1

Sharron H. Francis and Jackie D. Corbin
Sildenafil, pharmacology of a highly selective PDE5 inhibitor 15

Culley C. Carson III
Sildenafil citrate: a 5-year update on the worldwide treatment of
20 million men with erectile dysfunction 35

*Francesco Montorsi, Alberto Briganti, Andrea Salonia,
Patrizio Rigatti and Arthur L. Burnett*
Current and future strategies for preventing and managing erectile
dysfunction following radical prostatectomy 49

Michael Müntener and Brigitte Schurch
Neurologic erectile dysfunction 67

Matthias J. Müller
Therapy of erectile dysfunction (ED) with sildenafil improves
quality of life (QoL) and partnership (QoP) 83

Stuart N. Seidman
Erectile dysfunction, depression, and pharmacological treatments:
biologic interactions 101

Jennifer T. Anger and Jennifer R. Berman
Potential role for the PDE5 inhibitor sildenafil in the treatment of
female sexual dysfunction 117

Shadwan F. Alsafwah and Stuart D. Katz
Molecular processing of sildenafil in endothelial function: potential
applications in cardiovascular diseases 129

Graham Jackson
Sildenafil and cardiovascular events – drug interactions 143

Stephan Rosenkranz and Erland Erdmann
Cardiovascular safety of sildenafil in the treatment of
erectile dysfunction 151

Hossein Ardeschir Ghofrani, Werner Seeger and Friedrich Grimminger
NO pathway and phosphodiesterase inhibitors in pulmonary
arterial hypertension 163

Udo Dunzendorfer, Arne Behm, Eva Dunzendorfer and Annette Dunzendorfer
The phosphodiesterase V inhibitor low responder study (PILRS)
in patients with erectile dysfunction – A rationale for a PDE5
inhibitor combination therapy 169

Mirko Müller
Extended clinical use of sildenafil in patients with IPP, prostatitis and
infertility syndrome 183

Jan Dunzendorfer, Udo Dunzendorfer, Harald Förster
The cultural impact of sildenafil 187

Index 195

List of contributors

Shadwan F. Alsafwah, Department of Internal Medicine, Section of Cardiovascular Medicine, Yale University School of Medicine, 135 College Street, Suite 301, New Haven, CT 06510, USA; e-mail: salsafwah@msn.com

Jennifer T. Anger, UCLA Department of Urology, Box 951738, Los Angeles, CA 90095-1738, USA; e-mail: janger@mednet.ucla.edu

Arne Behm, Department of Urology, University of Lübeck Medical School, Ratzeburger Allee 160, 23538 Lübeck, Germany; e-mail: arne@behm.de

Jennifer R. Berman, The Female Sexual Medicine Center, UCLA Department of Urology, 924 Westwood Blvd., Suite 515, Los Angeles, CA 90033, USA; e-mail: jberman@mednet.ucla.edu

Alberto Briganti, Department of Urology, Universita' Vita-Salute San Raffaele, Via Olgettina 60, 20132 Milan, Italy; e-mail: briganti.alberto@hsr.it

Arthur L. Burnett, Department of Urology, The Johns Hopkins Hospital, Baltimore, Maryland 21287, USA; e-mail: aburnett@jhmi.edu

Culley C. Carson, Division of Urology, School of Medicine, University of North Carolina at Chapel Hill, 2140 Bioinformatics Building, Campus Box 7235, Chapel Hill, NC 27599-7235, USA; e-mail: carson@med.unc.edu

Jackie D. Corbin, Dept. of Molecular Physiology & Biophysics, Vanderbilt University School of Medicine, Light Hall Room 702, Nashville, TN 37232-0615, USA; e-mail: Jackie.corbin@vanderbilt.edu

Annette Dunzendorfer, Evangelisches Krankenhaus Ludwigsfelde-Teltow, Akademisches Lehrkrankenhaus der Freien Universität Berlin, Innere Medizin, Albert-Schweitzer-Str. 40–44, 14974 Ludwigsfelde, Germany; e-mail: annette@dunzendorfer.de

Eva Dunzendorfer, Alois Eckertstr. 28, 60528 Frankfurt/Main, Germany; e-mail: eva@dunzendorfer.de

Jan Dunzendorfer, Humboldt University Berlin, Kulturwissenschaftliches Institut, Sophienstr. 22a, 10178 Berlin, Germany; e-mail: jan@dunzendorfer.de

Udo Dunzendorfer, J.-W.-G.-University, Zeil 65–69, 60313 Frankfurt/Main, Germany; e-mail: dr.udo@dunzendorfer.de

Erland Erdmann, Klinik III für Innere Medizin, Universität zu Köln, Joseph-Stelzmann-Str. 9, 50924 Köln, Germany

Harald Förster, J.-W.-G.-University, Zeil 65–69, 60313 Frankfurt/Main, Germany

Sharron H. Francis, Dept. of Molecular Physiology & Biophysics, Vanderbilt University School of Medicine, Light Hall Room 702, Nashville, TN

37232-0615, USA; e-mail: Sharron.francis@vanderbilt.edu

Hossein Ardeschir Ghofrani, Department of Internal Medicine, Pulmonary Hypertension Center, University Hospital, Klinikstrasse 36, 35392 Giessen, Germany; e-mail: Ardeschir.Ghofrani@innere.med.uni-giessen.de

Friedrich Grimminger, Department of Internal Medicine, Pulmonary Hypertension Center, University Hospital, Klinikstrasse 36, 35392 Giessen, Germany; e-mail: friedrich.grimminger@innere.med.uni-giessen.de

Graham Jackson, Cardiothoracic Centre, St Thomas Hospital, Lambeth Palace Road, London SE1 7EH, UK; e-mail: graham@jacksonmd.fsnet.co.uk

Stuart D. Katz, Department of Internal Medicine, Section of Cardiovascular Medicine, Yale University School of Medicine, 135 College Street, Suite 301, New Haven, CT 06510, USA; e-mail: stuart.katz@yale.edu

Francesco Montorsi, Department of Urology, Universita' Vita-Salute San Raffaele, Via Olgettina 60, 20132 Milan, Italy; e-mail: montorsi.francesco@hsr.it

Mirko Müller, Department of Urology, Moorenstrasse 5, 40225 Düsseldorf, Germany; e-mail: muellemi@uni-duesseldorf.de

Matthias J. Müller, Department of Psychiatry, University of Mainz, Untere Zahlbacher Str. 8, 55131 Mainz, Germany; e-mail: mjm@mail.psychiatrie.klinik.uni-mainz.de

Michael Müntener, Department of Urology, University Hospital, Frauenklinikstr. 10, 8091 Zürich, Switzerland

Ian H. Osterloh, Pfizer Ltd, Ramsgate Road, Bldg. 500, IPC 475, Sandwich, Kent CT13 9NJ, United Kingdom; e-mail: Ian_Osterloh@sandwich.pfizer.com

Patrizio Rigatti, Department of Urology, Universita' Vita-Salute San Raffaele, Via Olgettina 60, 20132 Milan, Italy

Stephan Rosenkranz, Klinik III für Innere Medizin, Universität zu Köln, Joseph-Stelzmann-Str. 9, 50924 Köln, Germany; e-mail: stephan.rosenkranz@medizin.uni-koeln.de

Andrea Salonia, Department of Urology, Universita' Vita-Salute San Raffaele, Via Olgettina 60, 20132 Milan, Italy; e-mail:salonia.andrea@hsr.it

Brigitte Schurch, Spinal Cord Injury Center, University Hospital Balgrist, Forchstrasse 340, 8008 Zürich, Switzerland

Werner Seeger, Department of Internal Medicine, Pulmonary Hypertension Center, University Hospital, Klinikstrasse 36, 35392 Giessen, Germany; e-mail: werner.seeger@innere.med.uni-giessen.de

Stuart N. Seidman, Department of Psychiatry, College of Physicians and Surgeons of Columbia University, and the New York State Psychiatric Institute, 1051 Riverside Drive, Unit 98, New York, NY 10032, USA; e-mail: sns5@columbia.edu

Preface

In a series of text books on sildenafil this MDT volume is not the first, but it is certainly the most comprehensive guide to date. It provides information on biochemistry, physiology, pharmacology and also clinical experience accumulated over a number of years since sildenafil was first marketed. In addition it deals with its effects on civilization and culture in modern times, which may be summed up as mistrust changed to enthusiastic welcome. The blockbuster drug sildenafil has been intensely studied in a vast array of comorbidities and is currently awaiting the next generation of similar drugs in attempting to usurp its superior position in the market.

The drug is used daily in many different treatment regimens to challenge personal distress caused by a variety of diseases, helping to overcome both temporary and permanent loss of function. This volume also helps to demonstrate the handling of patients struggling to keep normalcy in their lives in difficult situations. Often the demands for treatment and the stage the research has reached do not go hand in hand; there is still a long way to go until the drug works under extrinsic as well as intrinsic conditions.

Udo Dunzendorfer   Frankfurt am Main, Mai 2004

The discovery and development of Viagra® (sildenafil citrate)

Ian H. Osterloh

Viagra Medical Strategies, Pfizer Ltd, Ramsgate Road, Bldg. 500, IPC 475, Sandwich, Kent, CT13 9NJ, UK

Introduction

Sildenafil citrate (UK-92,480, Viagra®) is a selective inhibitor of phosphodiesterase type 5 (PDE5) and acts on the nitric oxide (NO)/cyclic guanosine monophosphate (cGMP) pathway [1, 2]. UK-92,480 was first synthesised in the Sandwich laboratories of Pfizer Ltd, UK in 1989, and resulted from a discovery programme aimed at developing a selective inhibitor of PDE5. The origins of the project that eventually led to the discovery and development of Viagra date from around the mid 1980s. At this time, scientists working at the Pfizer European Research Centre were interested in potential new approaches to the treatment of several cardiovascular diseases.

Angina was an indication of particular interest. It was known that nitrates were effective agents for the short-term treatment of angina attacks, but the development of tolerance relatively quickly after initiation of therapy limited the clinical usefulness of nitrates for the chronic treatment of angina. It was already recognised that nitrate administration leads to the exogenous production of NO, a reactive gaseous molecule that has a number of actions [3]. With regard to antianginal effects, it was known that NO could diffuse across cell membranes and catalyse the enzyme guanylate cyclase to convert guanosine triphosphate (GTP) to cGMP. Production of cGMP within a vascular smooth muscle cell eventually leads to a lowering of intracellular calcium ions and smooth muscle relaxation. Thus, nitrates exert a therapeutic effect by causing smooth muscle relaxation; there may be a reduction in both preload and afterload for the heart, and possibly the patency of blood supply to the ischaemic myocardium is also improved [4].

There were also several theories as to why tolerance to nitrates develops quickly. Scientists at Pfizer hypothesised that a new treatment approach could be devised that prevented the natural breakdown of the second messenger cGMP, rather than increasing the production of NO. If this was possible, then an effective antianginal agent could be developed that might not be associated with tachyphylaxis during long-term therapy.

In the mid-1980s five families of phosphodiesterases (PDEs) were well recognised (Tab. 1) [5]. Some of these enzymes catalysed the breakdown of cyclic adenosine monophosphate (cAMP; PDE3, PDE4), some catalysed the breakdown of cGMP (PDE5), and some catalysed the breakdown of both cAMP and cGMP (PDE1, PDE2). In 1986, a project team was established, and the PDE5 enzyme was chosen as the most appropriate target for development of an inhibitor. The rationale was that PDE5 was known to be present in vascular smooth muscle and in platelets. Thus, an agent that selectively inhibited PDE5 should have vasodilatory and platelet antiaggregatory properties and, thus, have at least a dual mechanism of action for the treatment and prevention of angina attacks.

Although at the time it was not known whether it would be feasible to develop a selective inhibitor of PDE5 (devoid of action against PDEs 1–4), the project team succeeded in achieving the synthesis of novel pyrazolopyrimidines that were highly potent inhibitors of PDE5. Eventually, a compound designated chemically as 1-[[3-(6,7-dihydro-1-methyl-7-oxo-3-propyl-1H-pyrazolo[4,3-d]pyrimidin-5-yl)-4-ethoxyphenyl]sulfonyl]-4-methylpiperazine citrate was chosen for further profiling (Fig. 1) [6].

Table 1. Tissue distribution and cyclic nucleotide specificity of human PDEs [5, 10, 38, 68]

PDE Family	Substrate specificity	Tissue Localization	*Sildenafil IC_{50} (nM)	Selectivity Ratio
1	cGMP cAMP	Brain, heart, kidney, liver, skeletal muscle, vascular and visceral smooth muscle	280	80
2	cAMP cGMP	Adrenal cortex, brain, corpus cavernosum, heart, kidney, liver, visceral smooth muscle, and skeletal muscle	>30,000	>8,570
3	cAMP cGMP	Corpus cavernosum, heart, platelets, vascular and visceral smooth muscle, liver, fat, kidney	16,200	4,630
4	cAMP	Kidney, lung, mast cells, brain, heart, skeletal muscle, vascular and visceral smooth muscle, thyroid, testis	7,680	2,190
5	cGMP	Corpus cavernosum, platelets, vascular and visceral smooth muscle	3.5	—
6	cGMP	Retinal rod and cone cells	35	10
7	cAMP	Skeletal muscle, heart, lymphocytes	21,300	6,100
8	cAMP	Widely distributed; most abundant in testis, ovary, small intestine, colon	29,800	8,500
9	cGMP	Widely distributed; most abundant in spleen, small intestine, brain	2,610	750
10	cAMP cGMP	Putamen and caudate nucleus; testis, thyroid	9,800	2,800
11	cAMP cGMP	Corpus cavernosum, penile vasculature, smooth muscle, testis, pituitary, liver, kidney, prostate, heart	2,730	780

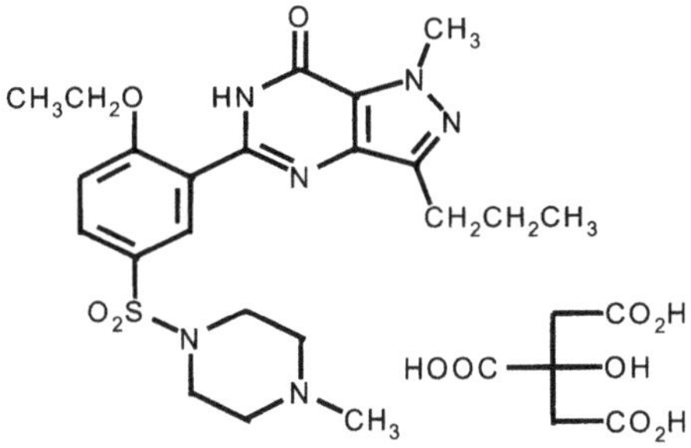

Figure 1. The chemical structure of Viagra (as the citrate salt)

The compound was demonstrated to have an IC_{50} of 3.5 nM for PDE5 and with excellent ratio of selectivity over PDEs 1–4 [2, 7]. The compound UK-92,480 (now known as sildenafil citrate) was recommended for full development in 1989 and entered clinical development in 1991 with the lead indication being angina. Thus, at this stage, it was hypothesised that sildenafil would be a mixed dilator of arteries and veins, similar to glyceryl trinitrate that also raises cGMP levels, and would also inhibit platelet aggregation. In addition, because sildenafil does not directly generate NO and acts downstream of NO [8], it was hypothesised that sildenafil would not be associated with any of the negative effects of directly raising levels of NO and associated free radicals. Therefore, it could be devoid of tachyphylaxis, which is a problem limiting the utility of nitrates in clinical practice. Preclinical studies were encouraging in confirming that sildenafil did have the properties of a mixed arteriovenous dilator and sildenafil was also found to have some antiplatelet aggregatory activity [9, 10].

Early clinical development

In early clinical studies performed in 1991 and 1992, sildenafil was found to have simple linear pharmacokinetics [11, 12], and also some vasodilatory activity, and was shown to modestly lower the blood pressure in healthy volunteers [9]. The magnitude of the vasodilatory effect was rather modest compared with that of nitrates alone. Moreover, sildenafil was found to interact with nitrates, with the combination leading to exaggerated decreases in blood pressure in some individuals [13]. Because nitrates are frequently prescribed to men with angina, further development of sildenafil for the lead indication of angina would have significant hurdles to overcome. Furthermore, sildenafil had a relatively short half-life, and when administered three times per day to healthy volunteers, it was associated with several adverse events. One of these adverse events, reported for the first time in Clinical Study 148-207, was penile erection. This was a study in healthy volunteers, and the design was routine for the pharmacokinetic profiling of an agent intended to be administered on a chronic basis. Young healthy male volunteers were admitted to a Phase I Clinical Unit and received three-times-a-day administration of placebo or various doses of sildenafil for 10 consecutive days. At higher doses some of the volunteers did report penile erections occurring more frequently or lasting longer than usual. Initially, the reports of erections as adverse events were not considered of special importance. Erections had not been reported in previous single dose studies, even with doses as high as 200 mg. Moreover, as mentioned earlier, the design of the first study in which erections were reported involved administration of sildenafil three times per day for 10 consecutive days. Erections were not reported on day one but only after several day's dosing and at dose regimens that produced other adverse events such as headache and indigestion. Furthermore, the subjects were young healthy male volunteers. Even if sildenafil did have an erectogenic effect, it was still a great extrapolation to assume that similar effects might occur in middle-aged and elderly men with erectile dysfunction (ED) secondary to vascular disease such as hypertension, cardiac disease, or diabetes. Finally, how many patients would be willing to start treatment three times a day, in the middle of the week, in order to have an erection at the weekend?

Nevertheless, it was this clinical observation that eventually led to a change in the direction of the preclinical and clinical research programmes. To make sense of the observations from the Phase I study, it had to be assumed that PDE5 was present in the smooth muscle of the corpus cavernosum, and this assumption was later confirmed by in-house *in vitro* experiments on samples of human penis [7]. The Pfizer project team members postulated that if PDE5 played a prominent role in corpus cavernosum smooth muscle physiology, then administration of sildenafil could enhance the erectile response to sexual stimulation, based on neurogenic and endothelial release of NO, while having little or no effect on the penis in the absence of sexual activity. During sexual activity, the cavernous nerves would release NO, which would diffuse across cell membranes into the smooth muscle of the corpus cavernosum to stimulate the production of cGMP, and the breakdown of cGMP by PDE5 would be inhibited by sildenafil. In the absence of sexual stimulation, there would be insufficient basal release of NO to form significant amounts of cGMP; thus, sildenafil would not be able to boost cGMP levels during periods of no sexual stimulation.

The Pfizer project scientists and clinicians obtained in-house approval to conduct a pilot clinical study in men with ED, but one immediate problem was how to measure an erectogenic effect of the drug if this effect occurred only during sexual stimulation. The clinical team members travelled to Southmeads Hospital, Bristol, UK to consult Mr Clive Gingell, a consultant urologist with a long track record of interest and research in ED. Eventually, the idea of monitoring erectile activity by means of the Rigiscan device was adopted. The Rigiscan apparatus contains two loops that are attached around the penis and can continuously monitor circumference and rigidity; the information can be transferred to a computer disc attached to the device and, later, to a remote personal computer. The advantage of using Rigiscan was that the patient would be able to view erotic videos in private, while objective evidence of the timing, magnitude, and duration of any erectile response could be obtained.

Thus, the first pilot study in ED (Study 148-350, which recruited 16 men with "psychogenic" ED) was performed in late 1993 (with Clive Gingell and Sam Gepi-Attee as investigators) and demonstrated that sildenafil could enhance erectile activity. However, in this particular study a three-times-a-day dose regimen had been used and the Rigiscan monitoring had not been introduced until the patient had had seven continuous days of therapy with sildenafil or placebo. Within the project team, opinion was divided over whether a single dose of sildenafil could also produce an erectogenic effect, but clearly the potential value of sildenafil would be much greater if it could be demonstrated to work soon after administration of a single dose. A second pilot study (148-351) performed in the first half of 1994 investigated whether single doses of placebo, sildenafil 10 mg, 25 mg, or 50 mg could be effective in producing erections during sexual stimulation [14]. The study recruited 12 men with "psychogenic" ED, and the results showed that these doses were well tolerated. Moreover, there was a very clear dose-response relationship in terms of duration of rigid erection [14]. Furthermore, in a follow-up phase to this study, the patients indicated success in using the drug at home during sexual relations with their partner.

Full clinical development

Although the drugs and devices for ED at that time were often effective (Tab. 2), only an estimated 10% of men sought treatment. Because of this, in the middle of 1994, there was real excitement among the project team that sildenafil could be a breakthrough in the treatment of ED. However, many questions remained. First, the two ED studies had in total recruited only 28 men, all from the same centre in Bristol. Could the results be replicated on a larger scale and at other centres? Also, none of the men had an obvious medical cause of the ED. Would sildenafil also work in men with diseased blood

Table 2. Pre-Viagra treatment options for erectile dysfunction

Year introduced	Name of treatment
1973	Penile implants
1973	Vascular surgery to correct venous leakage
1983	Vacuum pump
1995	Vasoactive intracavernosal pharmacotherapy
1997	Transurethral alprostadil, PGE_1

vessels, neurological, or other causes of ED? Should the trials recruit only men in whom the cause of ED had been identified? Should only certain types of patients with specified causes of ED be recruited? Would sildenafil be well tolerated when administered for several weeks or months continuously, and would efficacy be maintained or would tachyphylaxis occur? The results with Rigiscan monitoring had been impressive, but could sildenafil be shown to work in the home setting? There appeared to be no satisfactory diaries or questionnaires to evaluate efficacy in the home setting; could such instruments be developed and validated? Would the concept of treating ED and the use of diaries and questionnaires be acceptable to the regulators?

New members joined the project team and started working on these problems. It was decided that for pivotal trials, the effectiveness of the treatment would have to rely on reports by the patient (and the partner if willing). There should be no mechanical or other devices interrupting the natural setting of sexual activity. The team members decided to develop their own self-report diaries and questionnaires for patients and partners, and to work with external experts to refine and validate these new instruments. One of the outcomes of this work was the development and validation of the International Index of Erectile Function (IIEF), a 15-item self-report questionnaire [15]. The regulators accepted this instrument as appropriate for the assessment of efficacy, and it has become the gold standard efficacy assessment instrument for use in clinical trials of ED.

The programme was designed to test sildenafil in an increasing diversity of patients with a range of organic causes of ED [16], and to cover a wide rang of doses (5–200 mg). It soon became clear that sildenafil was having a dramatic effect on the sex lives of many of the trial patients. Normally, when a preregistration clinical trial ends, the patients return to standard therapy, but many of the patients and their partners were so grateful to have sexual relations restored that it was unacceptable to them to stop treatment. Eventually, the project team obtained agreement that every patient who completed a double-blind trial would be eligible to enrol in a long-term noncomparative, open-label sildenafil trial, and these trials would be continued until such time as regulatory approval was obtained in each country or until the programme had to be halted for unexpected reasons.

Also, the efficacy and toleration results were very promising [17]. When the project team first became aware that sildenafil was having a positive effect in patients with ED, many of us still considered that efficacy might be limited to relatively healthy men without overt cardiac disease, diabetes, or other medical risk factors. In fact, as the trials progressed encouraging data on the efficacy of sildenafil were generated from men with hypertension [18], heart disease [19, 20], diabetes [21], spinal cord injury [22, 23], in men who had undergone radical prostate cancer surgery [24, 25], and in elderly patients [26]. After registration, evidence for the efficacy of sildenafil would expand further to include men with Parkinson's disease [27], multiple sclerosis [28], severe renal failure [29, 30], kidney transplant [31], heart transplant [32, 33], heart failure [34], and depression [35].

Investigation of safety

The preregistration clinical trials programme progressed very quickly, but one minor unexpected adverse event reported at the highest doses was mild reversible visual effects including blue tinge to vision and increased perception of brightness of light [36]. As a result of this finding, potential visual effects of sildenafil were extensively investigated before regulatory approval [37].

Increasing members of the PDE family were being identified by Pfizer scientists and others (Tab. 1). Sildenafil inhibitor activity was characterized against these new PDEs [38]. Sildenafil had an approximately 10-fold selectivity for PDE5 *versus* PDE6, and much greater margins of selectivity, 2 orders of magnitude or greater, for PDEs 7–11. Thus, it was concluded that at therapeutic doses, sildenafil would have no clinically significant effect on PDEs 1–4 and 7–11, but in addition to effects on PDE5, it might have a clinically significant effect through the inhibition of PDE6.

Further *in vitro* and *in vivo* investigations confirmed that the transient visual side effects were associated with inhibition of PDE6, a cGMP-metabolising enzyme exclusively present in photoreceptors. Sildenafil, which has a similar potency *versus* PDE6 in all species tested for toxicology, has short-term reversible effects on electrical response to light. In patients, this is the likely explanation for the transient blue tinge to vision and increased perception of brightness. In preclinical studies long-term sustained inhibition of PDE6 by sildenafil, even at doses many multiples of those administered to humans, did not lead to any lasting damage to the structure and function of the eye [37]. Short- and long-term clinical studies also confirmed the transient nature of the visual side effects and the lack of long-term consequences [37].

All of the other adverse events could be related to the inhibition of PDE5 [39, 40]. Thus, headache, flushing, and nasal congestion were all considered to be related to vasodilatory effects. Dyspepsia was most likely related to inhibition of the smooth muscle of the oesophagogastric sphincter and perhaps an increased tendency to reflux.

Cardiovascular safety was extensively investigated pre- and postregistration with respect to haemodynamic studies, drug–drug interaction studies, and larger long-term studies in patients with various cardiovascular diseases [41]. The results are discussed in more detail in other Chapters, but in brief, sildenafil was shown to cause modest reductions in blood pressure in healthy volunteers [9]. Sildenafil could be administered safely to the vast majority of patients with ED, including those with stable heart disease or taking most classes of antihypertensives [20, 42], but it should not be administered to patients receiving nitrates because of the risk of an exaggerated hypotensive response in susceptible individuals [43].

Viagra enters the market

By the time that Pfizer was in a position to file a registration dossier, more than 4,500 male subjects had been exposed to sildenafil, and 21 separate efficacy studies had been completed; all of these studies showed positive results. Viagra was approved in many major markets in 1998 (Tab. 3), and it was first launched in the United States in April 1998 (with a starting dose of 50 mg and the option to increase the dose to 100 mg or decrease to 25 mg, depending on efficacy and toleration) [44]. The launch of Viagra revolutionized the management of ED, but more important, the efficacy and safety experience in the clinic closely matched that reported from clinical trials [45–48]. Viagra was found to be well tolerated and effective in a very large proportion of patients (Fig. 2) [49, 50], and at the time of preparing this Chapter, more than 20 million men have received prescriptions for Viagra.

The success of Viagra has stimulated the development of other PDE5 inhibitors, and two new ones have entered the market in Europe. It is excellent news for patients that there are now more treatment options than a few years ago. However from the data available, Viagra has unsurpassed efficacy and safety and is likely to remain the gold standard treatment for ED [39, 51].

Table 3. Development of Viagra (sildenafil citrate) timeline

Year	Event
1989	Pfizer scientists synthesise sildenafil citrate
1991	Two researchers at Pfizer Central Research in Sandwich, England – Peter Ellis and Nick Terrett – note that drugs capable of inhibiting phosphodiesterase type 5 (PDE5) might be helpful in treating erectile dysfunction (ED).
1991	Early Phase I studies of sildenafil citrate for treating angina – single-dose studies involving healthy volunteers – yield no findings of note.
1992	Second Phase I angina study – a multiple-dose study on healthy volunteers – reveals that erections are a side effect of using the drug.
1992	First and only Phase II clinical trial of sildenafil citrate as a treatment for angina finds mild hemodynamic effects.
Late 1993	First pilot study on sildenafil citrate as a treatment for ED is carried out in Bristol, England. Men took the drug three times a day for a week.
Early 1994	Second pilot study on sildenafil as an ED treatment shows that a single dose is capable of enabling erections.
1994	Pfizer accelerates its research pace on sildenafil citrate as a treatment for ED.
1994	Pfizer scientists detect the enzyme PDE5 in corpus cavernosal tissue, confirming that sildenafil's mechanism of action in treating ED is by inhibiting PDE5.
1994–1997	21 clinical trials on sildenafil citrate, now known as Viagra, involve nearly 4500 men with ED.
1997	Pfizer files a New Drug Application for Viagra with the US Food and Drug Administration (FDA). The agency gives Viagra priority review status, which is reserved for drugs that represent major advances in treatment or fulfil a significant medical need.
March 1998	US FDA approves Viagra as the first oral medication for the treatment of ED.

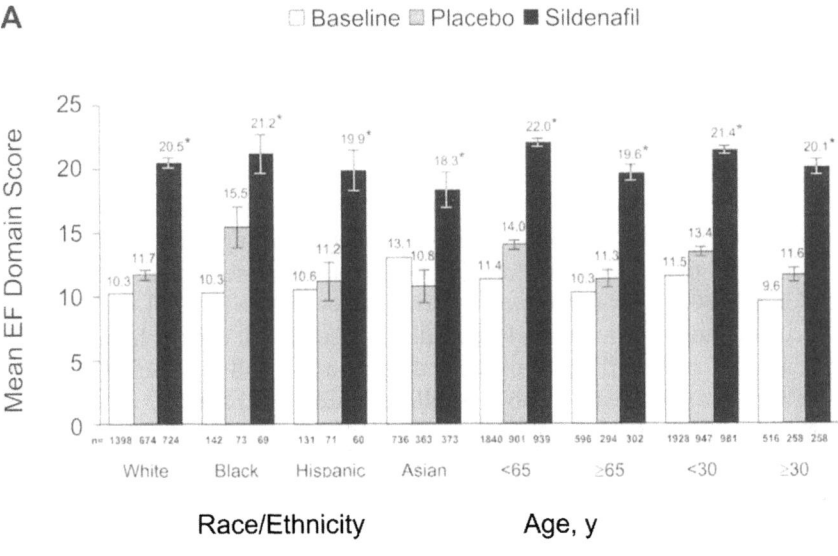
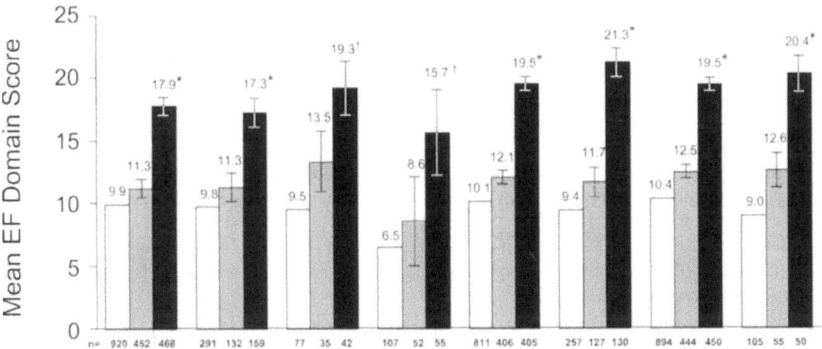

Figure 2. Overall baseline mean score and least squares mean (±) SE scores at end of treatment for the erectile function (EF) domain (questions 1–5 and 15) of the International Index of Erectile Function from 11 pooled double-blind, placebo-controlled clinical trials. Total domain score range is 1 to 30; scores for individual questions range from 1 (never/almost never) to 5 (always/almost always), with a response of 0 indicating "did not attempt sexual intercourse". Panel A shows scores for subgroups characterised by patient characteristics. Panel B shows scores for subgroups characterised by concomitant medical conditions/medications. anti-HTNs = antihypertensives, BMI = body mass index, HTN = hypertension, IHD = ischemic heart disease, PVD = peripheral vascular disease, RP = radical prostatectomy. $*P < 0.0001$, $^{\dagger}P < 0.001$. Reprinted with permission from *Urology* (2002), 60 (Suppl 2B): 18–27 [51]

Future developments

Viagra is currently being investigated for its potential role in various female sexual disorders, and it is to be hoped that some female sexual disorders will be amenable to treatment with sildenafil, perhaps combined with other forms of treatment in some cases [52–55]. Viagra originally entered clinical development as a potential treatment for angina, so it is a very interesting story that there is now renewed interest in the use of sildenafil to treat various cardiovascular conditions [56–59]. At the time of this writing, Viagra shows great potential for treating various forms of pulmonary hypertension in adults and in children [60–64], for treating congestive cardiac failure [65, 66], and for other vascular conditions including Raynaud's disease [67]. It is to be hoped that the potential of sildenafil in all of these disease areas will be confirmed by the results of prospective, placebo-controlled, double-blind studies.

References

1 Jeremy JY, Ballard SA, Naylor AM, Miller MAW, Angelini GD (1997) Effects of sildenafil, a type-5 cGMP phosphodiesterase inhibitor, and papaverine on cyclic GMP and cyclic AMP levels in the rabbit corpus cavernosum *in vitro*. *Br J Urol* 79: 958–963
2 Boolell M, Allen MJ, Ballard SA, Gepi-Attee S, Muirhead GJ, Naylor AM, Osterloh IH, Gingell C (1996) Sildenafil: an orally active type 5 cyclic GMP-specific phosphodiesterase inhibitor for the treatment of penile erectile dysfunction. *Int J Impot Res* 8: 47–52
3 Moncada S, Higgs A (1993) The L-arginine-nitric oxide pathway. *N Engl J Med* 329: 2002–2012
4 Hare JM, Colucci WS (1995) Role of nitric oxide in the regulation of myocardial function. *Prog Cardiovasc Dis* 38: 155–166
5 Beavo JA (1995) Cyclic nucleotide phosphodiesterases: functional implications of multiple isoforms. *Physiol Rev* 75: 725–748
6 Terrett N, Bell A, Brown D, Ellis P (1996) Sildenafil (Viagra), a potent and selective inhibitor of type 5 cGMP phosphodiesterase with utility for the treatment of male erectile dysfunction. *Bioorg Med Chem Lett* 6: 1819–1824
7 Ballard SA, Gingell CJ, Tang K, Turner LA, Price ME, Naylor AM (1998) Effects of sildenafil on the relaxation of human corpus cavernosum tissue *in vitro* and on the activities of cyclic nucleotide phosphodiesterase isozymes. *J Urol* 159: 2164–2171
8 Lue TF (2000) Erectile dysfunction. *N Engl J Med* 342: 1802–1813
9 Jackson G, Benjamin N, Jackson N, Allen MJ (1999) Effects of sildenafil citrate on human hemodynamics. *Am J Cardiol* 83: 13C–20C
10 Wallis RM, Corbin JD, Francis SH, Ellis P (1999) Tissue distribution of phosphodiesterase families and the effects of sildenafil on tissue cyclic nucleotides, platelet function, and the contractile responses of trabeculae carneae and aortic rings *in vitro*. *Am J Cardiol* 83: 3C–12C
11 Muirhead GJ, Rance DJ, Walker DK, Wastall P (2002) Comparative human pharmacokinetics and metabolism of single-dose oral and intravenous sildenafil. *Br J Clin Pharmacol* 53 (Suppl 1): 13S–20S
12 Nichols DJ, Muirhead GJ, Harness JA (2002) Pharmacokinetics of sildenafil citrate after single oral doses in healthy male subjects: absolute bioavailability, food effects and dose proportionality. *Br J Clin Pharmacol* 53: 5S–12S
13 Webb DJ, Freestone S, Allen MJ, Muirhead GJ (1999) Sildenafil citrate and blood-pressure-lowering drugs: results of drug interaction studies with an organic nitrate and a calcium antagonist. *Am J Cardiol* 83: 21C–28C
14 Boolell M, Gepi-Attee S, Gingell JC, Allen MJ (1996) Sildenafil, a novel effective oral therapy for male erectile dysfunction. *Br J Urol* 78: 257–261

15 Rosen RC, Riley A, Wagner G, Osterloh IH, Kirkpatrick J, Mishra A (1997) The International Index of Erectile Function (IIEF): a multidimensional scale for assessment of erectile dysfunction. *Urology* 49: 822–830
16 Montorsi F, McDermott TE, Morgan R, Olsson A, Schultz A, Kirkeby HJ, Osterloh IH (1999) Efficacy and safety of fixed-dose oral sildenafil in the treatment of erectile dysfunction of various etiologies. *Urology* 53: 1011–1018
17 Dinsmore WW, Hodges M, Hargreaves C, Osterloh IH, Smith MD, Rosen RC (1999) Sildenafil citrate (VIAGRA) in erectile dysfunction: near normalization in men with broad-spectrum erectile dysfunction compared with age-matched healthy control subjects. *Urology* 53: 800–805
18 Kloner RA, Brown M, Prisant LM, Collins M (2000) Efficacy and safety of Viagra (sildenafil citrate) in patients with erectile dysfunction taking concomitant antihypertensive therapy. *Am J Hypertens* 14: 70–73
19 Conti CR, Pepine CJ, Sweeney M (1999) Efficacy and safety of sildenafil citrate in the treatment of erectile dysfunction in patients with ischemic heart disease. *Am J Cardiol* 83: 29C–34C
20 Olsson AM, Persson CA (2001) Efficacy and safety of sildenafil citrate for the treatment of erectile dysfunction in men with cardiovascular disease. *Int J Clin Pract* 55: 171–176
21 Rendell MS, Rajfer J, Wicker PA, Smith MD, for the Sildenafil Diabetes Study Group (1999) Sildenafil for treatment of erectile dysfunction in men with diabetes. *JAMA* 281: 421–426
22 Giuliano F, Hultling C, El Masry WS, Smith MD, Osterloh IH, Orr M, Maytom M (1999) Randomized trial of sildenafil for the treatment of erectile dysfunction in spinal cord injury. *Ann Neurol* 46: 15–21
23 Derry FA, Dinsmore WW, Fraser M, Gardner BP, Glass CA, Maytom MC, Smith MD (1998) Efficacy and safety of oral sildenafil (VIAGRA) in men with erectile dysfunction caused by spinal cord injury. *Neurology* 51: 1629–1633
24 Zippe CD, Kedia AW, Kedia K, Nelson DR, Agarwal A (1998) Treatment of erectile dysfunction after radical prostatectomy with sildenafil citrate (Viagra). *Urology* 52: 963–966
25 Lowentritt BH, Scardino PT, Miles BJ, Orejuela FJ, Schatte EC, Slawin KM, Elliott SP, Kim ED (1999) Sildenafil citrate after radical retropubic prostatectomy. *J Urol* 162: 1614–1617
26 Wagner G, Montorsi F, Auerbach S, Collins M (2001) Sildenafil citrate (VIAGRA) improves erectile function in elderly patients with erectile dysfunction: a subgroup analysis. *J Gerontol A Biol Sci Med Sci* 56: M113–M119
27 Hussain IF, Brady CM, Swinn MJ, Mathias CJ, Fowler CJ (2001) Treatment of erectile dysfunction with sildenafil citrate (Viagra) in parkinsonism due to Parkinson's disease or multiple system atrophy with observations on orthostatic hypotension. *J Neurol Neurosurg Psychiatry* 71: 371–374
28 Fowler C, Miller J, Sharief M (1999) Viagra (sildenafil citrate) for the treatment of erectile dysfunction in men with multiple sclerosis. *Ann Neurol* 46: 497
29 Chen J, Mabjeesh NJ, Greenstein A, Nadu A, Matzkin H (2001) Clinical efficacy of sildenafil in patients on chronic dialysis. *J Urol* 165: 819–821
30 Rosas SE, Wasserstein A, Kobrin S, Feldman HI (2001) Preliminary observations of sildenafil treatment for erectile dysfunction in dialysis patients. *Am J Kidney Dis* 37: 134–137
31 Prieto Castro RM, Anglada Curado FJ, Regueiro Lopez JC, Leva Vallejo ME, Molina Sanchez J, Saceda Lopez JL, Requena Tapia MJ (2001) Treatment with sildenafil citrate in renal transplant patients with erectile dysfunction. *BJU Int* 88: 241–243
32 Chait J, Kobashigawa J, Chuang J, Moriguchi J, Kawata N, Laks H (1999) Efficacy and safety of sildenafil citrate (Viagra) in male heart transplant patients. *J Heart Lung Transplant* 18: 58
33 Wagoner LE, Giesting RM, Bell BJ, McGuire NC, Abraham WT (1999) Is Viagra (sildenafil) safe and effective in cardiac transplant recipients? *Transplantation* 67: S102
34 Webster L, Michelakis E, Davis T, Tsuyuki R, Archer S (2002) Sildenafil is safe and effective treatment for erectile dysfunction in males with NYHA class II-III congestive heart failure. *Circulation* 106 (Suppl. 2): II-469
35 Seidman SN, Roose SP, Menza MA, Shabsigh R, Rosen RC (2001) Treatment of erectile dysfunction in men with depressive symptoms: results of a placebo-controlled trial with sildenafil citrate. *Am J Psychiatry* 158: 1623–1630
36 Morales A, Gingell C, Collins M, Wicker PA, Osterloh IH (1998) Clinical safety of oral sildenafil citrate (VIAGRA) in the treatment of erectile dysfunction. *Int J Impot Res* 10: 69–74
37 Laties AM, Zrenner E (2002) Viagra® (sildenafil citrate) and ophthalmology. *Prog Retin Eye Res* 21: 485–506

38 Corbin JD, Francis SH (2002) Pharmacology of phosphodiesterase-5 inhibitors. *Int J Clin Pract* 56: 453–459
39 Padma-Nathan H, Eardley I, Kloner RA, Laties AM, Montorsi F (2002) A 4-year update on the safety of sildenafil citrate (Viagra®). *Urology* 60 (Suppl. 2B): 67–90
40 Sadovsky R, Miller T, Moskowitz M, Hackett G (2001) Three-year update of sildenafil citrate (Viagra) efficacy and safety. *Int J Clin Pract* 55: 115–128
41 Shakir SAW, Wilton LV, Boshier A, Layton D, Heeley E (2001) Cardiovascular events in users of sildenafil: results from first phase of prescription event monitoring in England. *BMJ* 322: 651–652
42 Zusman R, Collins M (1999) Effect of sildenafil on blood pressure in men with erectile dysfunction taking concomitant antihypertensive medications. *J Am Coll Cardiol* 33: 238A
43 Zusman RM, Morales A, Glasser DB, Osterloh IH (1999) Overall cardiovascular profile of sildenafil citrate. *Am J Cardiol* 83: 35C–44C
44 Goldstein I, Lue TF, Padma-Nathan H, Rosen RC, Steers WD, Wicker PA (1998) Oral sildenafil in the treatment of erectile dysfunction. *N Engl J Med* 338: 1397–1404
45 Fagelman E, Fagelman A, Shabsigh R (2001) Efficacy, safety, and use of sildenafil in urologic practice. *Urology* 57: 1141–1144
46 Marks LS, Duda C, Dorey FJ, Macairan ML, Santos PB (1999) Treatment of erectile dysfunction with sildenafil. *Urology* 53: 19–24
47 Jarow JP, Burnett AL, Geringer AM (1999) Clinical efficacy of sildenafil citrate based on etiology and response to prior treatment. *J Urol* 162: 722–725
48 Moreira SG, Brannigan RE, Spitz A, Orejuela FJ, Lipshultz LI, Kim ED (2000) Side-effect profile of sildenafil citrate (Viagra) in clinical practice. *Urology* 56: 474–476
49 Moore R, Edwards J, McQuay H (2002) Sildenafil (Viagra) for male erectile dysfunction: a meta-analysis of clinical trial reports. *BMC Urol* 2: 1–12
50 Fink HA, Mac Donald R, Rutks IR, Nelson DB, Wilt TJ (2002) Sildenafil for male erectile dysfunction: a systematic review and meta-analysis. *Arch Intern Med* 162: 1349–1360
51 Carson CC, Burnett AL, Levine LA, Nehra A (2002) The efficacy of sildenafil citrate (Viagra®) in clinical populations: an update. *Urology* 60 (Suppl 2B): 12–27
52 Berman JR, Berman LA, Toler SM, Gill J, Haughie S for the Sildenafil Study Group (2003) Safety and efficacy of sildenafil citrate for the treatment of female sexual arousal disorder: a double-blind, placebo-controlled study. *J Urol* 170: 2333–2338
53 Caruso S, Intelisano G, Lupo L, Agnello C (2001) Premenopausal women affected by sexual arousal disorder treated with sildenafil: a double-blind, cross-over, placebo-controlled study. *Br J Obs Gynaecol* 108: 623–628
54 Nurnberg H, Hensley P, Lauriello J, Parker L, Keith S (1999) Sildenafil for women patients with antidepressant-induced sexual dysfunction. *Psychiatr Serv* 50: 1076–1078
55 Mark A, Shifren J (2003) Medical therapy for female sexual dysfunction. *Prim Care Update Ob/Gyns* 10: 40–43
56 Reffelmann T, Kloner RA (2003) Therapeutic potential of phosphodiesterase 5 inhibition for cardiovascular disease. *Circulation* 108: 239–244
57 Zhang R, Wang Y, Zhang L, Zhang Z, Tsang W, Lu M, Zhang L, Chopp M (2002) Sildenafil (Viagra) induces neurogenesis and promotes functional recovery after stroke in rats. *Stroke* 33: 2675–2680
58 Vlachopoulos C, O'Rourke MF, Hirata K (2001) Sildenafil (Viagra®) improves the elastic properties of the aorta. *Am J Hypertens* 14: 6A
59 Mahmud A, Hennessy M, Feely J (2001) Effect of sildenafil on blood pressure and arterial wave reflection in treated hypertensive men. *J Hum Hypertens* 15: 707–713
60 Lepore JJ, Maroo A, Pereira NL, Ginns LC, Dec GW, Zapol WM, Bloch KD, Semigran MJ (2002) Effect of sildenafil on the acute pulmonary vasodilator response to inhaled nitric oxide in adults with primary pulmonary hypertension. *Am J Cardiol* 90: 677–680
61 Watanabe H, Ohashi K, Takeuchi K, Yamashita K, Yokoyama T, Tran QK, Satoh H, Terada H, Ohashi H, Hayashi H (2002) Sildenafil for primary and secondary pulmonary hypertension. *Clin Pharmacol Ther* 71: 398–402
62 Abrams D, Schulze-Neick I, Magee A (2000) Sildenafil as a selective pulmonary vasodilator in childhood primary pulmonary hypertension. *Heart* 82: E4
63 Carroll WD, Dhillon R (2003) Sildenafil as a treatment for pulmonary hypertension. *Arch Dis Child* 88: 827–828
64 Ghofrani HA, Rose F, Schermuly RT, Olschewski H, Wiedemann R, Kreckel A, Weissmann N,

Ghofrani S, Enke B, Seeger W, Grimminger F (2003) Oral sildenafil as long-term adjunct therapy to inhaled iloprost in severe pulmonary arterial hypertension. *J Am Coll Cardiol* 42: 158–164
65 Hirata K, Adji A, Vlachopoulos C, O'Rourke MF (2003) Sildenafil improves left ventricular systolic function in patients with congestive heart failure: the role of wave reflections. *J Am Coll Cardiol* 41 (Suppl A): 264A
66 Katz SD (2003) Potential role of type 5 phosphodiesterase inhibition in the treatment of congestive heart failure. *Congest Heart Fail* 9: 9–15
67 Lichtenstein JR (2003) Use of sildenafil citrate in Raynaud's phenomenon: comment on the article by Thompson et al. *Arthritis Rheum* 48: 282–283
68 Francis SH, Turko IV, Corbin JD (2001) Cyclic nucleotide phosphodiesterases: relating structure and function. *Prog Nucleic Acid Res Mol Bio* 65: 1–52

Sildenafil, pharmacology of a highly selective PDE5 inhibitor

Sharron H. Francis and Jackie D. Corbin

Light Hall Room 702, Department of Molecular Physiology & Biophysics, Vanderbilt University School of Medicine, Nashville, TN 37232-0615, USA

Introduction

The roles of cAMP (cyclic AMP) and cGMP (cyclic GMP) in regulating myriad physiological and pathophysiological processes are well established, and our understanding of the actions of these nucleotides continues to rapidly expand. Cyclic AMP was discovered in the late 1950s by Dr. Earl Sutherland; the discovery of this novel compound derived from his studies on the regulation of glycogen metabolism. The vast number of biological systems affected by cAMP action was quickly appreciated [1]. Cyclic GMP in biological tissues was discovered somewhat later in 1963 [2]. However, aside from the role of cGMP in visual transduction, scientists had difficulty in establishing a definitive role for cGMP in physiological processes. The importance of cGMP as a second messenger in relaxation of vascular smooth muscle emerged in the 1970s, and today we realize the impact of this molecule in many other systems [3–5]. The role of cGMP in smooth muscle is the basis for the efficacious actions of nitrovasodilators such as nitroglycerin in the treatment of angina pectoris or cyclic nucleotide phosphodiesterase (PDE) inhibitors in the treatment of male erectile dysfunction (MED).

Dr. Earl Sutherland and his colleagues were the first to identify and characterize the adenylyl and guanylyl cyclases that synthesize cAMP and cGMP and the PDEs that degrade these compounds (Fig. 1). The phosphohydrolase activity associated with PDEs is highly specific for action against the novel six-member cyclic phosphate ring, which is the distinguishing feature of cyclic nucleotides; cAMP and cGMP are largely resistant to other phosphohydrolases [1]. While there are numerous adaptive processes for reducing the synthesis of cAMP and cGMP, the role of PDE action is absolutely critical in controlling cellular levels of these highly active signaling molecules because the catalytic action of PDEs provides essentially the only way for a cell to rapidly and efficiently lower its cyclic nucleotide content [6–9]. Furthermore, the catalytic activities of these enzymes are rapidly regulated by a variety of mechanisms in response to the changing intracellular milieu. A number of cells have been

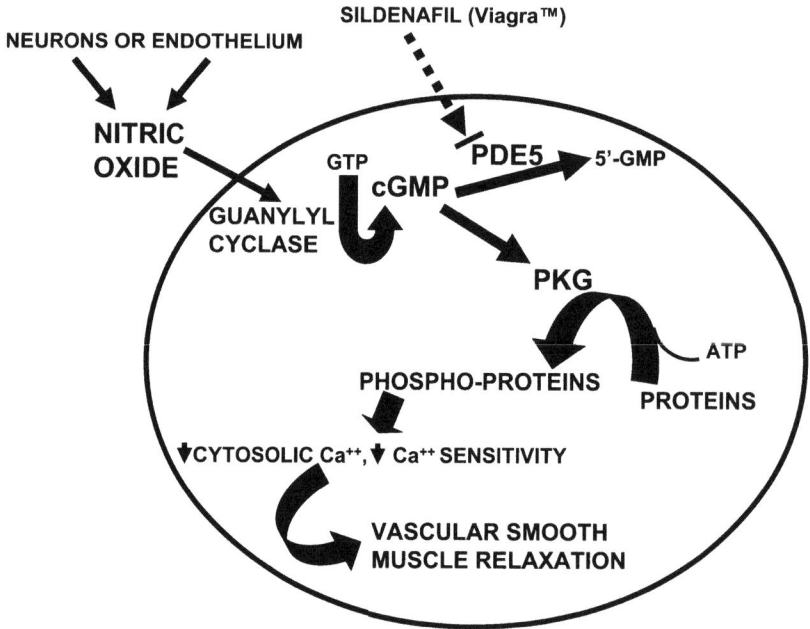

Figure 1. Regulation of smooth muscle relaxation in the penile corpora cavernosa and the effect of sildenafil to enhance cGMP accumulation following sexual arousal.

shown to extrude cyclic nucleotides into the extracellular space through the action of anion transporters, but on a quantitative basis, the contribution of these pumps to lowering cellular cyclic nucleotide content is low [10, 11].

Sutherland and colleagues also identified the first known PDE inhibitors, i.e., caffeine, theophylline, and related compounds. In the initial phases of research on cyclic nucleotides, agents such as these were very useful since they inhibit almost all known PDEs although they are relatively weak in potency. However, these early observations provided the biochemical basis for the development of more specific and potent PDE inhibitors which continues today and now provides for a number of clinical interventional therapies that specifically target particular PDEs. The pioneering work of these early investigators also established the synergistic action of PDE inhibitors with agents that activate the cyclases that synthesize cyclic nucleotides [1]. In the presence of a cyclase agonist alone, the increase in cellular cyclic nucleotide and the corresponding physiological effect is typically modest; however, when a PDE inhibitor is combined with agonist, there is a robust increase in both the cyclic nucleotide content of the cell and the physiological response. This basic concept of synergistic action (more than additive) has been exploited as a tool in thousands of studies of cyclic nucleotide action in cells, tissues, and whole organisms.

Despite the fact that caffeine and theophylline have been used clinically, there are significant limitations with these drugs. Caffeine and related natural compounds such as theophylline are non-specific PDE inhibitors, i.e., they inhibit the catalytic activity of almost all PDEs. This increases the likelihood for these compounds to affect multiple organs and physiological processes. They are also not entirely specific for members of the PDE family since they interact with non-PDE proteins; these latter interactions can affect a range of metabolic processes and are not trivial. Lastly, they are weak inhibitors requiring high dosages to achieve a response.

Mammalian PDEs belong to a superfamily of enzymes comprised of eleven families of PDEs (PDEs 1–11); each family has varying degrees of specificity to degrade cAMP and/or cGMP [6–8]. Many naturally occurring PDE inhibitors such as caffeine and its homologs inhibit most of these diverse families of PDEs. For decades, many non-selective PDE inhibitors, as well as synthetic PDE inhibitors, have been used to investigate physiological effects of cyclic nucleotides and PDE activity. However, the challenge to medicinal chemists, pharmacologists, and physiologists to identify compounds that are highly selective for particular PDEs has proved to be a daunting task. Literally thousands of synthetic compounds have been developed and tested for selectivity and potency against various PDEs, and there have been remarkable advances in improving selectivity for particular PDEs. Many of these compounds have been investigated for potential therapeutic efficacy. As a group, however, most of these inhibitors lack sufficient potency. Most also lack specificity because they have been to block the catalytic activity of several PDEs, including that of phosphodiesterase-5 (PDE5).

The advent of specific and potent PDE5 inhibitors

The persistence of several research groups has now led to the successful development and production of a number of compounds that are highly selective and potent for inhibition of specific PDEs and three of these to date have proved useful in treatment of male erectile dysfunction (MED) [12–24]. In the mid-1980s, zaprinast became available to the scientific community and its high selectivity for PDE5 and some isoforms of PDE1 was a significant step forward in developing a specific PDE5 inhibitor (Fig. 2). Subsequently, compounds with structures related to zaprinast including sildenafil (Viagra™) and vardenafil (Levitra™) were developed and proved to be much more potent and selective for PDE5. All three of these inhibitors share structural similarities with cGMP (Fig. 2). The development and successful marketing of the highly selective PDE5 inhibitor, sildenafil (Viagra™) by Pfizer, Inc. has reenergized the area of PDE inhibitor research. Sildenafil is highly selective for PDE5, and it is the first orally administered PDE5 inhibitor developed for the treatment of MED. A number of other orally administered PDE5 inhibitors with variations in potency, selectivity, and kinetic properties are currently being marketed in most countries and in the US or are approaching approval [16–24]. The use of sildenafil for treatment of MED has been validated by its clinical efficacy in

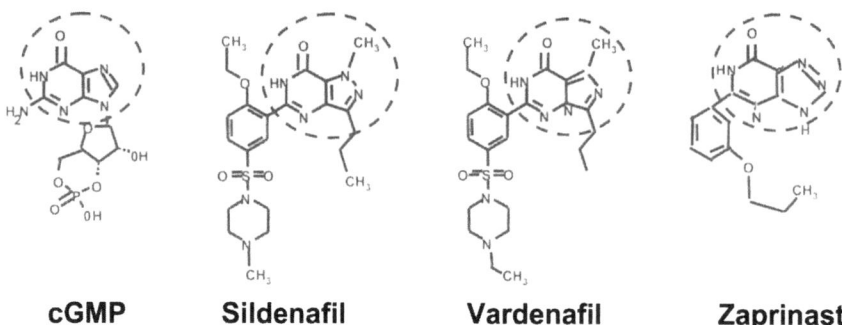

Figure 2. Comparison of the molecular structures of cGMP, sildenafil, vardenafil, and zaprinast. Circled areas indicate the ring structures in sildenafil, vardenafil, and zaprinast that resemble the purine moiety in cGMP.

improving erectile function in a large percentage of men with erectile dysfunction, and its safety profile is excellent [16–18]. The overall safety of Viagra™ in a variety of patient populations has been validated in more than 120 clinical trials, and despite more than five years of prescribing to more than 20 million patients worldwide, reported side effects are minimal. Whether the two other PDE5 inhibitors, i.e., vardenafil (Levitra™) and tadalafil (Cialis™), that have recently entered the marketplace will have the same safety and efficacy profile as sildenafil remains to be seen. Like sildenafil, tadalafil and vardenafil are highly selective and potent for the catalytic site of PDE5.

Since its introduction, sildenafil (Viagra™) has dominated the market for treatment of MED [12–18], and new prospects for its use in the treatment of other maladies are emerging [25–29]. The remarkable therapeutic success and safety of sildenafil has focused attention on better defining the properties of this and other PDE5 inhibitors as well as on the properties of PDE5 that contribute to the high affinity of these inhibitors for interaction with the PDE5 catalytic site. Reports indicate that sildenafil may be therapeutically efficacious in the clinical management of other disease processes including improved regulation of pulmonary artery pressure associated with pulmonary hypertension, relief of symptoms associated with Raynaud's phenomenon, improved recovery from stroke, and reduced tone in the lower esophageal sphincter of patients with achalasia [25–29]. Sildenafil may also be neuroprotective [30, 31]. Features of PDE5 that contribute to efficacy and potency of sildenafil and the pharmacokinetics of sildenafil will be discussed in this Chapter.

The clinical success and safety profile of sildenafil (Viagra™) is now very well documented [32–40]. The high potency and selectivity of sildenafil for PDE5 are certainly major factors in its efficacy and safety [15, 16], but other factors also contribute to its success; this includes the limited tissue distribution of PDE5 and the stringent specificity of PDE5 for cGMP as a substrate. The tissue distribution of PDE5 is restricted compared to other PDE families.

PDE5 is highly abundant in platelets, some neuronal cells, and notably in smooth muscle throughout the body including the vascular smooth muscle of the systemic vasculature and that of specialized vascular beds such as that of the penile corpora cavernosa [8, 13, 14, 41, 42]. Because of this, pharmacological intervention that targets PDE5 is less likely to have broad effects involving many tissues. As will be discussed below, specific targeting by PDE5 inhibitors is also achieved in part by the synergistic action of these drugs with sexual arousal. Importantly, PDE5 is not abundant in cardiac tissue, but it is undoubtedly present in the smooth muscle of the blood vessels traversing the heart muscle. Since PDE5 is highly specific for hydrolyzing cGMP, a PDE5 inhibitor is unlikely to elicit major problems in cAMP-signaling pathways. In some tissues containing the dual-specificity PDE3, blocking PDE5 action can also cause an elevation of cAMP [43]; this is because cGMP and cAMP are both hydrolyzed by PDE3, and they compete for the PDE3 catalytic site. Selective elevation of either of the nucleotides can decrease hydrolysis of the other. However, since activation of either the cAMP- and cGMP-signaling pathways brings about smooth muscle relaxation, the overall effect of the PDE5 inhibitor would not be impaired in that tissue. PDE3 is much more widely distributed than is PDE5. The likelihood of such "cross-talk" involving PDE5 in myocardial contractile cells is also diminished by the presence of low levels of PDE5 in these cells. Therefore, it is unlikely that PDE5 inhibitors would affect heart contractility by this mechanism.

Role of cGMP signaling in effects of PDE5 inhibitors

The role of cGMP in modulating function in smooth muscle throughout the body including that of the penile vasculature is well established; all share a common pathway for cGMP signaling through activation of cGMP-dependent protein kinase (PKG) (Fig. 1) [3–5]. Nitric oxide is released as a neurotransmitter from nerves innervating vascular structures in the penis, and it is also released as a paracrine agonist from endothelial cells lining blood vessel walls; nitric oxide initiates the cGMP-signaling cascade by diffusing into the vascular smooth muscle cell where it forms a high-affinity complex with a heme prosthetic group in the nitric oxide-sensitive guanylyl cyclase (GC) [44, 45]. When nitric oxide binds to the heme group, GC is activated, and production of cGMP from GTP is dramatically increased [44]. One of the main intracellular targets for cGMP is PKG; cGMP binds to allosteric cGMP-binding sites on PKG and activates this kinase [3–5]. The activated PKG then phosphorylates a number of proteins to cause diminished intracellular Ca^{2+} effects due to increased extrusion of Ca^{2+}, increased sequestration of Ca^{2+} in intracellular structures, and decreased sensitivity of target proteins to Ca^{2+} [3–5]. This results in decreased vasomotor tone and smooth muscle relaxation in all known smooth muscle cells. This process occurs selectively in the penile corpora cavernosa following sexual arousal due to increased local release of the neurotransmitter nitric oxide from the non-adrenergic, non-cholinergic nerve endings in the penis and from the arterial endothelial cells of the penile vasculature [45, 46]. As a result, blood flow into the penis increases, the capacitance of the corporeal sinusoids is enhanced, and blood accumulates within these structures. This provides for increased tumescence and an erectile response.

Modulation of cellular cGMP level

In all cells, the level of cGMP is primarily determined by the relative rates of cGMP synthesis by the GCs and its breakdown by PDEs (Fig. 1). An imbalance between the synthesis (e.g., decreased release of nitric oxide due to penile nerve deterioration or endothelial cell damage) and breakdown of cGMP can compromise the accumulation of cGMP in cells. Cyclic GMP-dependent protein kinase (PKG), the main mediator of smooth muscle relaxation, may become deficient, or intracellular Ca^{2+} could become excessive or overly active. Finally, among other insufficiencies, the contractile machinery itself may not be amenable to adequate relaxation. Under these scenarios, the cGMP-signaling pathway may not be sufficiently activated to elicit adequate vasodilation to foster penile erection. For example, the cGMP level achieved by sexual arousal may be insufficient to activate PKG and bring about proper relaxation of the penile vascular smooth muscle; this can result in MED. Since PDE5 is the major cGMP-hydrolyzing PDE in the penile corpora cavernosa, it is a logical target for pharmacological intervention to enhance cGMP accumulation by blocking the cGMP hydrolytic activity of this enzyme [41, 42]. Sildenafil alone has no effect on vascular smooth muscle tone, and in the absence of sexual arousal, it has no effect to facilitate penile erection. The need for the driving force of the nitric oxide agonist in order for sildenafil to act suggests that basal GC activity is low in the penile vasculature. However, when PDE5 catalytic activity is blocked by an inhibitor such as sildenafil, even a modest elevation of GC activity could produce a robust elevation of intracellular cGMP thereby improving erectile function. This is consistent with the action of nitric oxide and sildenafil on separate steps within the same nitric oxide/cGMP signaling pathway, with nitric oxide increasing cGMP formation by GC and sildenafil efficiently blocking cGMP breakdown [8, 13, 14]. This synergistic effect amplifies cGMP accumulation and provides at least in part for targeting of the sildenafil effect to the penis. This also the reason that concomitant use of nitrovasodilators (which release nitric oxide) and sildenafil is strongly contraindicated since severe systemic hypotension may result [47].

Mechanism of action of sildenafil

Sildenafil is a competitive and reversible inhibitor of cGMP hydrolysis by the catalytic site of PDE5. Sildenafil is an analog of cGMP, the PDE5 substrate, but PDE5 has ~1000-fold higher affinity for sildenafil than for cGMP (Fig. 2) [13–15]. The purine ring of cGMP is mimicked by similar components in sildenafil, and this undoubtedly provides in part for the specificity of the compound for the PDE5 catalytic site (Fig. 2). However, the novel components of the sildenafil structure provide for the higher affinity, and these additional elements in the sildenafil structure most likely exploit interactions in and around the PDE5 catalytic site that are not utilized in interacting with cGMP. Furthermore, these distinctive structural components most likely contribute to the specificity of the compound for the PDE5 catalytic site since the structure cannot be accommodated in other cGMP-binding sites such as those found in PKG and cGMP-gated cation channels, in the catalytic sites of other PDEs, or in the allosteric cGMP-binding sites of PDE5 [48]. The characteristics of the novel interactions between sildenafil and the catalytic site of PDE5 are emphasized by the fact that the potency of sildenafil can be significantly altered by minor changes in its molecular structure. Although the molecular differences in sildenafil and vardenafil are apparently modest, vardenafil is ~10–40 times more potent in inhibiting PDE5 [15]. Among several possibilities, this could be due to the different positions of nitrogen atoms in the imidazotriazinone ring of the two compounds, which might engender differences in electron distribution.

Structure and function of PDE5

PDE5 is a dimeric enzyme that is composed of two identical 100 kD proteins (Fig. 3) [8, 14, 49]. There are four known isoforms of PDE5 (PDE5A 1–4) [50–52] ACC.No: NM_033431; these isoforms are products of a single gene and are formed by alternative splicing of mRNA. The enzymatic characteristics of the PDE5A1–3 appear to be quite similar although there may be some selectivity in their tissue distributions; PDE5A3 is largely expressed in smooth muscle [52]. The enzymatic properties of PDE5A4 have not been characterized. Each of the monomers in PDE5 is a chimeric protein that contains two major functional domains that are approximately equal in size, i.e., a catalytic domain (C domain) located in the more C-terminal portion of the protein and a regulatory domain (R domain) located in the more N-terminal portion [8, 14, 49]. The single catalytic site located in the C domain is the target for sildenafil [12–16]. The R domain of PDE5 contains allosteric cGMP-binding sites that contribute importantly to regulation of enzyme functions and to potency of these PDE5 inhibitors. However, the allosteric cGMP-binding sites in PDE5 are evolutionarily and biochemically distinct from that of the catalytic site. These sites are highly specific for cGMP; and do not interact with sildenafil [48, 49, 53]. The catalytic site binds cGMP in a shallow pocket along the surface of the enzyme. When cGMP occupies this site, the cyclic phosphate bond of cGMP is brought into proximity with the catalytic machinery of the enzyme, which involves an array of amino acids and divalent cations includ-

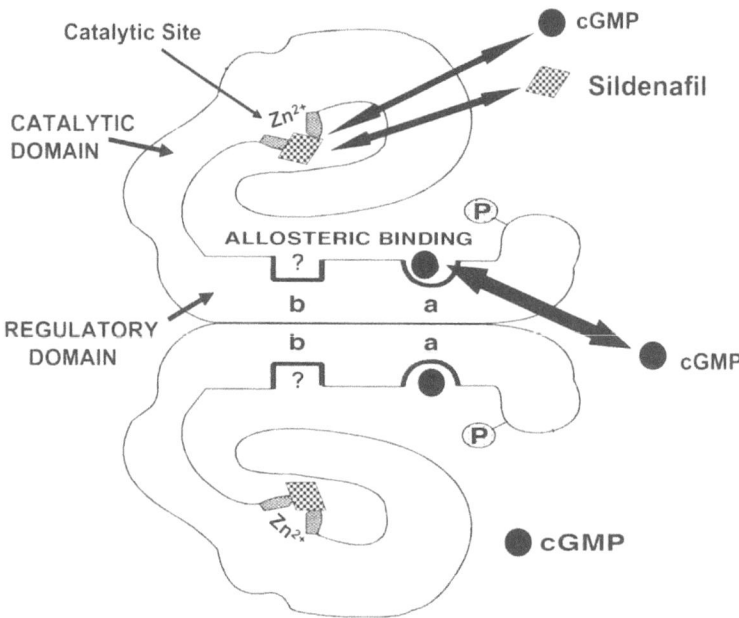

Figure 3. Working model of PDE5. PDE5 is a dimer of two identical monomers. Each monomer contains a catalytic domain and a regulatory domain. Sildenafil blocks catalysis by binding to the catalytic site and competing with the substrate, cGMP. Cyclic GMP binds to allosteric sites in the regulatory domain.

ing Zn^{2+} [8, 14]. This arrangement provides for the rapid hydrolysis of the cyclic phosphate bond of cGMP to form 5'-GMP, which has low affinity for the enzyme and rapidly dissociates from PDE5. 5'-GMP is inactive in the cellular cGMP-signaling pathway. Other cellular phosphohydrolases do not hydrolyze the novel cyclic phosphate bond of cGMP or cAMP. Because the structure of sildenafil resembles that of cGMP (Fig. 2), it can occupy the PDE5 catalytic site, thus blocking access to cGMP. In addition, sildenafil is stable and is not inactivated by the catalytic machinery; nor is it metabolized significantly in the smooth muscle cell. For these reasons, occupation of the PDE5 catalytic site by sildenafil competitively inhibits cGMP breakdown since cGMP cannot gain access to the catalytic machinery. In the face of ongoing synthesis of cGMP in any tissue containing PDE5, this will cause cGMP to accumulate and to increase cGMP signaling through PKG. In the penile corpora cavernosa, this contributes to improved erectile function.

Although the C domain of PDE5 is the direct target of PDE5 inhibitors (Fig. 3), functions of the R domain enhance the PDE5 inhibitor actions on the enzyme. Allosteric cGMP-binding is provided by sites in the R domain; whether one or two cGMP molecules are bound per subunit is still unclear. In addition, there is a single consensus phosphorylation site for PKG or PKA near the N-terminus. Phosphorylation of this site activates PDE5 catalytic function and thereby provides for negative feedback regulation of cGMP levels [54–56]. Phosphorylation of this serine is tightly controlled by cGMP levels since occupation of the PDE5 allosteric cGMP-binding sites is required for phosphorylation to occur, and the site is preferentially phosphorylated by PKG compared to PKA. Therefore, it is likely that the site is only phosphorylated when cGMP is elevated in the cell. When cGMP binds to the allosteric sites, cGMP is not degraded as it is in the catalytic site, but PDE5 enzyme functions are altered. Cyclic GMP binding to the allosteric sites in the R domain produces a conformational change that exposes the serine allowing it to be rapidly phosphorylated, thereby increasing PDE5 catalytic activity [8, 14, 54–56]. Cyclic GMP occupation of the allosteric sites also increases the affinity of the catalytic site for cGMP, thereby further activating PDE5 catalytic site functions [57, 58]. However, in the presence of a PDE5 inhibitor and ongoing synthesis of cGMP, cellular cGMP is elevated, which fosters increased binding of cGMP to the allosteric sites and phosphorylation of the serine by activated PKG; as a result, the affinity with which PDE5 inhibitors bind at the catalytic site is increased. Therefore, due to its molecular mechanism, the potency of sildenafil is actually greater than would occur in the absence of the R domain. This property of the enzyme translates into greater clinical efficacy and potency of sildenafil and other PDE5 inhibitors. Following ingestion of a PDE5 inhibitor tablet, any elevation of cGMP in smooth muscle cells should increase the avidity with which the PDE5 catalytic site binds that inhibitor. That is, the PDE5 inhibitor, by fostering increased binding of cGMP to PDE5 allosteric sites, stimulates its own efficacy and potency (Fig. 1) [59].

Assessing the potencies of PDE5 inhibitors

Assuming that all other factors are equal, the higher the biochemical affinity (potency) of an inhibitor for PDE5, the lower the expected dose of the inhibitor that will be needed [15]. This concept of potency can be quantitatively assessed *in vitro* by determining the concentration of a particular inhibitor that inhibits PDE5 activity by 50%; this value is referred to as the IC_{50}, which reflects the affinity of the protein for that molecule. Efficacious PDE5 inhibitors should have affinities, i.e., IC_{50} values, in the nanomolar (nM) range. However, pharmacokinetics for each compound strongly impact the dose required in patients. Higher potency *in vitro* does not mean that an inhibitor will have a greater clinical effect, but that it is likely that less of it will be needed to achieve the desired result.

According to literature values and our own determinations, the *in vitro* biochemical potency of sildenafil is in the low nM range [13–16]. The IC_{50} values reported in the literature vary somewhat (1–9 nM), which most likely reflects differences in assay conditions and enzyme sources. In our laboratory, we have used the purified recombinant human PDE5 assayed at 0.4 µM cGMP as substrate to determine the IC_{50} value for sildenafil to be 3.7 ± 1.4 nM. When patients are administered a 100-mg dose of sildenafil, the free concentration of sildenafil in the blood reaches ~40 nM [60–62]; if this fully equilibrates across the cell membrane, it should be more than sufficient to block PDE5 catalytic activity as long as the intracellular concentration of PDE5 is within this same concentration range.

In addition to classical IC_{50}, there are other methods to measure the potency of a drug and to examine whether a PDE5 inhibitor binds to a single site or multiple sites on the protein. In our laboratory, we have recently used radiolabeled sildenafil to directly measure the affinity of binding to PDE5, the stoichiometry of sildenafil binding per PDE5 molecule, and the biochemical properties that affect its binding by PDE5 [59]. The value obtained by this method reflects binding affinity and is termed K_D instead of IC_{50}. Compared with IC_{50}, the measurement of a K_D value is a more direct method of determining potency since it does not depend on competition with cGMP to block ongoing enzyme activity. The K_D of a PDE5 inhibitor obtained using this approach should numerically approach the IC_{50} of that inhibitor if the inhibitor only interacts with the catalytic site. In fact, our results show that the K_D value for sildenafil approximates its IC_{50} value. The agreement between these two values suggests that sildenafil does not bind to an appreciable extent to sites on PDE5 other than the C domain. This is also supported by the fact that the stoichiometry of binding to PDE5 is typically 1 mol/mol subunit or less.

The approach of determining the K_D value not only provides a powerful tool to measure PDE inhibitor potency but has already revealed unknown characteristics associated with the interaction between PDE5 and PDE5 inhibitors such as sildenafil. Most of these have not been possible to study previously. These include the rates of association to and dissociation from the PDE5 for the PDE5 inhibitors, information which could assist in predicting the rapidity of onset and the duration of inhibitor action in patients. The *in vitro* potency of a PDE5 inhibitor is not the same as efficacy. As discussed below, efficacy is based on the actual *in vivo* (clinical), effects of the inhibitor. Radiolabeled sildenafil can also be used to search for other proteins that might bind this compound. If other proteins, not necessarily PDEs, are found to bind an inhibitor such as sildenafil with high affinity, this could increase the vigilance of investigators and clinicians with regard to potential side effects of a particular PDE5 inhibitor in a given tissue.

Selectivity of PDE5 inhibitors

Assuming that a particular PDE5 inhibitor does not interact with other cellular proteins, the biochemical selectivity of that compound for PDE5 *versus* other PDEs is a key factor in determining its side effect profile [15]. When the relative affinity (IC_{50}) of an inhibitor for PDE5 *versus* its affinity for other PDEs (or other proteins) is sufficiently different, it is less likely that at therapeutic doses the compound will achieve plasma concentrations that will allow interaction with these non-target proteins. For a PDE5 inhibitor, selectivity is usually measured in terms of potency (IC_{50}) to inhibit PDE5 compared to inhibition of others members of the PDE superfamily (Tab. 1). This value is computed by dividing the IC_{50}s of the two compounds that are being compared. The superfamily of mammalian PDEs is comprised of 11 families (PDEs 1–11) of enzymes that catalyze the termination of cAMP and/or cGMP activity in cells by breaking the phosphodiester bond of these compounds. The activities associated with each of these PDE families are known or are implicated in the regulation of a broad range of cellular functions. There are at least 20 genes for mammalian PDEs since some of the eleven PDE families are derived from multiple genes, and some of these 20 genes have multiple products brought about by alternative mRNA splicing, resulting in more than 50 PDEs.

Table 1. Selectivity of sildenafil for inhibition of PDE5 compared to other PDEs [15]

PDE family	Fold selectivity	Substrate specificity
PDE1	80	cAMP & cGMP
PDE2	>19,000	cAMP & cGMP
PDE3	4600	cAMP & cGMP
PDE4	2100	cAMP
PDE5	1	cGMP
PDE6	10	cGMP
PDE7	6100	cAMP
PDE8	8500	cAMP
PDE9	750	cAMP & cGMP
PDE10	2800	cAMP & cGMP
PDE11	780	cAMP & cGMP

Like PDE5, the retinal PDEs, i.e, the PDE6 family, is also specific for cGMP, and the members of this PDE family are closely related to PDE5 both in structure and in biochemical properties. Thus, it is not surprising that sildenafil cross-reacts slightly with these PDEs. The IC_{50} value for sildenafil interaction with PDE5 is approximately 3–10 fold lower than that for the cone and rod isoforms of PDE6, respectively [13–15]. This may account for the transient dose-related impairment of blue/green color discrimination experienced by some patients (~3% of patients) taking sildenafil [62–64]; this brief effect coincides with the peak plasma concentration of sildenafil and is fully reversible. Sildenafil does not cross-react significantly with any PDE other than PDE6, i.e., the IC_{50} of sildenafil for PDE5 is more than 1000 times lower than those for most of the other PDEs [15]. The incidence of other reported side effects of sildenafil is low [headaches (12%), dyspepsia (5%), rhinitis (2%), flushing (9%), and diarrhea (2%), etc.]; some patients also experience a transient and slight lowering of blood pressure (5–8 mM Hg). Most of these side effects are consistent with inhibition of PDE5 and increased cGMP signaling in smooth muscle tissues outside the penile corpus cavernosum. While sildenafil has been reported to interact with some members of the multi-drug anion transporter protein, the affinity of this interaction is low, so it seems unlikely that this family of proteins contributes to the effects of sildenafil.

Safety of sildenafil in clinical use

The potential for adverse cardiovascular effects of the PDE5 inhibitors such as sildenafil has been an ongoing concern for clinicians. However, the results of literally hundreds of studies overwhelmingly support the conclusion that when used as recommended, this agent is safe in a broad range of patients, including those with active cardiovascular disease [16–18, 32–40]. This is supported by the paucity of reports of adverse events in the millions of men who have taken sildenafil since its introduction into the market in 1998 and in the low percentage of patients discontinuing its use due to adverse events. However, the use of this medication should always be medically supervised since the patient's medical history and current medication regimen must be considered before determining the advisability of using sildenafil and deciding on the dose that is best suited for an individual's needs. Use of sildenafil by patients being treated with nitrovasodilators (e.g., nitroglycerin) either on a regular or intermittent basis is strongly contraindicated due to the synergistic effect of sildenafil and the nitric oxide derived from the nitrovasodilator on smooth muscle relaxation [47]. This warning is based on the fact that there are no clear data on when the effects of the nitric oxide are fully dissipated from tissues. In addition, some patients with underlying medical conditions can be particularly sensitive to the actions of any vasodilator, including sildenafil [16]. The concomitant use of sildenafil and other medications that could delay serum clearance of this drug (e.g., the HIV protease inhibitor, ritonavir; erythromycin; other cytochrome P450 inhibitors; or cimetidine derivatives) can increase the serum concentrations of sildenafil dramatically [65, 66]; Elderly individuals also have slower drug clearance rates. For these patients, use of sildenafil should be closely monitored for side effects, and it is recommended that patients should initially use the lowest dosage.

In normal volunteers, sildenafil has little or no effect on cardiac index, cardiac output, systemic blood pressure, or platelet aggregation [16–18, 32–36]. Furthermore, it appears to be safe for patients tanking antihypertensive medication [16, 37]; there is no synergism with the action of antihypertensive mediations such as Ca^{2+}-channel blockers, α-adrenoreceptor or β-adrenoreceptor blockers, ACE inhibitors, or diuretics [16]. However, simultaneous administration of sildenafil and α-blockers may lead to symptomatic hypotension in some patients. For this reason, lower dosage of sildenafil is indicated for patients using α-blockers. Despite the presence of PDE5 in platelets and the anti-aggregatory role of the cGMP-signaling pathway in these cells, use of sildenafil does not significantly affect bleeding time [16, 67, 68]. Sildenafil does not increase the hypotensive effects of alcohol, and it does not interact with the effects of warfarin. A modest effect of sildenafil on platelet aggregation occurs only at the highest recommended dose of sildenafil. When taken in conjunction with aspirin, there is no further increase in bleeding time. Incidence of other cardiovascular events, including stroke, is not increased, and in an animal model, sildenafil administration following an ischaemic episode actually was efficacious in that it reduced infarct volume and improved recovery from the associated neurological deficits [31]. Whether this effect was due only to the vasodilatory effects of sildenafil is not known since the results of other studies indicate that sildenafil is protective for cultured neuronal cells [30].

Some have suggested that in platelets and in smooth muscle, the effect of sildenafil to block PDE5 and elevate cGMP could cause activation of the cAMP-signaling pathway because cGMP competes well with cAMP interaction at the catalytic site of another PDE in these tissues, i.e., PDE3, a dual-specificity PDE [43]. However, there is no measurable increase in cAMP in tissues treated with sildenafil. Likewise, the effect of sildenafil on systemic blood pressure is minor (5–10 mm Hg) even though both PDE5 and PDE3 are present in vascular smooth muscle. Therefore, it seems unlikely that sildenafil acts through elevation of cAMP. PDE3 is abundant in the heart, but there is low PDE5 in myocardial cells, and there is little effect of sildenafil on the heart. Another dual-specificity PDE, PDE2, is abundant in the heart, but cGMP activates this enzyme to cause increased hydrolysis of cAMP; if sildenafil increased cGMP in the heart, the effect on this enzyme would tend to lower cAMP and thereby decrease cardiac intropy [8]. The body of evidence derived from both clinical studies and basic research is entirely consistent with the interpretation that PDE5 inhibitors act by elevating cGMP and thereby activating the cGMP-signaling pathway through PKG. When used as recommended, these inhibitors are not associated with significant adverse cardiovascular events.

Pharmacokinetic properties of sildenafil

In addition to the biochemical properties discussed above, the pharmacokinetic properties of the PDE5 inhibitors (absorption, movement in the circulation, bioavailability, tissue uptake, elimination) undoubtedly have an impact on efficacy [15–18, 60–62]. There are several common pharmacokinetic parameters that can be measured and quantified that describe bodily distribution of a PDE5 inhibitor (Tab. 2). Bioavailability, maximum plasma concentration (C_{max}), time (T_{max}) required for attaining C_{max}, and time ($t_{1/2}$) required for elimination of one-half of the inhibitor from plasma are all important factors. Pharmacokinetic properties of a PDE5 inhibitor influence the time of onset as well as duration of its effect on penile erection.

Table 2. Pharmacokinetic parameters of sildenafil[1]

Parameter	Sildenafil[2] (fasted) 100 mg
C_{max}, ng/mL	450
t_{max}, hr	0.8
$t_{1/2}$, hr	3–5
Plasma protein binding	96%
Bioavailability (% ingested)	40%

[1] Absolute values may vary depending on the age of the patient, co- administration of other medications, and ingestion with a heavy meal.
[2] Viagra USPI [package insert]. 2000

Factors that contribute to estimating the appropriate range for the therapeutic dose

As noted above, a number of factors including the medical history of the patient and current medications are major considerations when initiating use of any new medication. Issues that impact the safety and efficacy of sildenafil for a given patient include the bioavailability of the drug, rate of clearance, and its interaction with other drugs. Sildenafil is dispensed in 25, 50, and 100 mg tablets. The bioavailability is the percentage of ingested inhibitor that actually appears in the plasma; approximately 40% of the sildenafil is absorbed and available to the tissues (Tab. 2). Of this 96% is reversibly bound to plasma proteins, and ~4% is unbound [15, 62–64]. Based on studies using a 100-mg oral dose, the peak plasma concentration of the unbound, active sildenafil is typically ~40 nM. The T_{max} varies from 0.5–2 h following ingestion with some variation due to differences among patients and differences in the circumstances under which the medication is ingested [15, 62, 63], and the erectogenic effect of sildenafil coincides with the T_{max}. For example, absorption of sildenafil is significantly slowed if ingested following a heavy meal, particularly if the meal has a high fat content, and the C_{max} can be reduced by as much as 30% [16]. The C_{max} value after a 100-mg oral dose of sildenafil in normal individuals is 450 ng/ml, but in certain patient populations, the C_{max} may be much higher, and the area under the curve (AUC) may also differ. This includes patients on certain medications such as protease inhibitors, cimetidine, erhythromycin and ketoconazole (see above) [65, 66], patients with hepatic impairment, and elderly patients [69]. For example, in a study comparing elderly male patients (65–81 years) with young male volunteers, the C_{max} was 60–70% higher in the older patients, and there was a significant elevation of the AUC for the older individuals [16, 69]. These factors increase the drug exposure of the patient and the frequency of use of sildenafil may need to be restricted. As with all medications, chronic or excessive use of sildenafil could increase the risk of side effects.

Clearance of PDE5 inhibitors from the plasma and from cells

Sildenafil and its metabolites are primarily excreted as metabolites in the feces (80%) and urine (13%) [62]. Sildenafil is metabolized in the liver primarily by four cytochromes, including CYP3A4, CYP2C9, CYP2C19, AND CYP2D6, and caution should be used when considering co-administration of sildenafil and drugs that inhibit these enzymes [70, 71]. A significant portion of the circulating sildenafil is modified in the liver to a major circulating metabolite (UK-103-320). This metabolite represents approximately 40% of the circulating sildenafil and is 2-fold lower potency for PDE5 than is sildenafil. It is estimated that ~20% of the pharmacological effect of sildenafil is due to this metabolite. A high proportion (96%) of both sildenafil and its metabolite are bound to proteins in the circulation.

Variation in the clearance time for sildenafil in different patient populations is also an important consideration. In healthy young volunteers, the $t_{1/2}$ of clearance of sildenafil from the plasma is ~4 hours, and at 16 hours virtually no sildenafil can be detected in the plasma. However, clearance from the plasma of older patients can be much delayed; in one study of elderly patients, sildenafil was still present at detectable levels in the plasma 36 h after taking the medication.

Since sildenafil is apparently not degraded by PDE5 or any other enzyme in smooth muscle cells, decline in PDE5 inhibitor action in cells may be slower than that predicted by the disappearance of the drug from the plasma. To exit the cell, the inhibitor must dissociate from its high-affinity site on PDE5, diffuse through the cytosol to the plasma membrane without being rebound by PDE5, traverse the plasma membrane to exit the cell, and then be transported to the liver via the bloodstream before it can be degraded. These factors undoubtedly affect the rate of exit of a PDE5 inhibitor from smooth muscle cells. This emphasizes the importance of avoiding even intermittent use of nitrovasodilators with sildenafil since it is difficult to know exactly when the drug is completely cleared from vascular smooth muscle cells. In fact, there are numerous anecdotal reports that the effect of sildenafil in relieving MED can occur well outside the time required for plasma clearance; if the effect is still present, an effective concentration of the drug must still be present in penile smooth muscle. For some medical uses currently being considered for PDE5 inhibitors, slow clearance from cells could prove to be advantageous, while for other treatments, it may be undesirable. However, it is an issue that must be considered since studies of clearance of PDE5 inhibitors from plasma are well documented, but studies of clearance of these inhibitors from smooth muscle cells are lacking.

Conclusion

The commercial availability of sildenafil (Viagra™), a highly selective and potent PDE5 inhibitor for use in therapeutic treatment of MED, is an exciting outcome of many years of research and pharmacological testing. The emerging prospect for use of sildenafil and newer PDE5 inhibitors in treatment of other maladies emphasizes the need to more fully understand the actions of these inhibitors and the biological systems in which they act. A number of pharmaceutical companies are currently developing other PDE5 inhibitors with novel structures and pharmacokinetics. New methodologies such as the use of radiolabeled PDE5 inhibitors to directly characterize the interaction of PDE5 inhibitors with PDE5 are providing new information that has already improved our understanding of how PDE5 inhibitors function, and this approach is likely to be helpful in predicting efficacy of new PDE5 inhibitors.

Acknowledgements
Supported by NIH grants DK40029 and DK58277.

References

1. Robison GA, Butcher RW, Sutherland EW (1971) *Cyclic AMP*. Academic Press, New York
2. Ashman DF, Lipton R, Milicow MM, Price TD (1963) Isolation of cAMP and cGMP from rat urine. *Biochem Biophys Res Comm* 11: 330–334
3. Francis SH, Corbin JD (1999) Cyclic nucleotide-dependent protein kinases: intracellular receptors for cAMP and cGMP action. *Crit Rev Clin Lab Sci* 36: 275–328
4. Lincoln TM, Dey N, Sellak H (2001) cGMP-dependent protein kinase signaling mechanisms in smooth muscle: from the regulation of tone to gene expression. *J Appl Physiol* 91: 1421–1430
5. Schlossmann J, Feil R, Hofmann F (2003) Signaling through NO and cGMP-dependent protein kinases. *Ann Med* 35: 21–27
6. Burns F, Zhao AZ, Beavo JA (1996) Cyclic nucleotide phosphodiesterases: gene complexity, regulation by phosphorylation, and physiological implications. *Adv Pharmacol* 36: 29–48
7. Soderling SH, Beavo JA (2000) Regulation of cAMP and cGMP signaling: new phosphodiesterases and new functions. *Curr Opin Cell Biol* 12: 174–179
8. Francis SH, Turko IV, Corbin JD (2001) Cyclic nucleotide phosphodiesterases: relating structure and function. *Prog Nucleic Acid Res Mol Biol* 65: 1–52
9. Rybalkin SD, Yan C, Bornfeldt KE, Beavo JA (2003) Cyclic GMP phosphodiesterases and regulation of smooth muscle function. *Circ Res* 93: 280–301
10. Mercapide J, Santiago E, Alberdi E, Martinez-Irujo JJ (1999) Contribution of phosphodiesterase isoenzymes and cyclic nucleotide efflux to the regulation of cyclic GMP levels in aortic smooth muscle cells. *Biochem Pharmacol* 58: 1676–1683
11. Ahlstrom M, Lamberg-Allardt C (1999) Regulation of adenosine 3',5'-cyclic monophosphate (cAMP) accumulation in UMR-106 osteoblast-like cells: role of cAMP-phosphodiesterase and cAMP efflux. *Biochem Pharmacol* 58: 1335–1340
12. Jeremy JY, Ballard SA, Naylor AM, Miller MA, Angelini GD (1997) Effects of sildenafil, a type-5 cGMP phosphodiesterase inhibitor, and papaverine on cyclic GMP and cyclic AMP levels in the rabbit corpus cavernosum *in vitro*. *Br J Urol* 79: 958–963
13. Ballard SA, Gingell CJ, Tang K, Turner LA, Price ME, Naylor AM (1998) Effects of sildenafil on the relaxation of human corpus cavernosum tissue *in vitro* and on the activities of cyclic nucleotide phosphodiesterase isozymes. *J Urol* 159: 2164–2171
14. Corbin JD, Francis SH (1999) Cyclic GMP phosphodiesterase 5: target for sildenafil. *J Biol Chem* 274: 13729–13732
15. Corbin JD, Francis SH (2002) Pharmacology of phosphodiesterase-5 inhibitors. *Int J Clin Pract* 56: 453–459
16. Rotella DP (2002) Phosphodiesterase 5 inhibitors: current status and potential applications. *Nat Rev Drug Discov* 1: 674–682
17. Salonia A, Rigatti P, Montorsi F (2003) Sildenafil in erectile dysfunction: a critical review. *Curr Med Res and Opin* 19: 241–262

18 Montorsi F, Salonia A, Deho F, Cestari A, Guazzoni G, Rigatti P, Stief C (2003) Pharmacological management of erectile dysfunction. *Brit J Urol* 91: 446–454
19 Hellstrom WJ, Gittelman M, Karlin G, Segerson T, Thibonnier M, Taylor T, Padma-Nathan H for the Vardenafil Study Group (2003) Sustained efficacy and tolerability of vardenafil, a highly potent selective phosphodiesterase type 5 inhibitor, in men with erectile dysfunction: results of a randomized, double-blind, 26-week placebo-controlled pivotal trial. *Urology* 61: 8–14
20 Klotz T, Sachse R, Heidrich A, Jockenhovel F, Rohde G, Wensing G, Horstmann R, Engelemann R (2001) Vardenafil increases penile rigidity and tumescence in erectile dysfunction patients: a RigiScan and pharmacokinetic study. *World J Urol* 19: 32–39
21 Lue TF (2000) Erectile dysfunction. *N Engl J Med* 324: 1801–1813
22 Porst H, Padma-Nathan H, Giuliano F, Anglin G, Varanese L, Rosen R (2003) Efficacy of tadalafil for the treatment of erectile dysfunction at 24 and 36 hours after dosing: a randomized controlled trial. *Urology* 62: 121–125
23 Pomerol JM, Rabasseda X (2003) Tadalafil, a further innovation in the treatment of sexual dysfunction. *Drugs Today* 39: 103–113
24 Hellstrom WJ, Gittelman M, Karlin G, Segerson T, Thibonnier M, Taylor T, Padma-Nathan H (2003) Sustained efficacy and tolerability of vardenafil, a highly potent selective phosphodiesterase type 5 inhibitor, in men with erectile dysfunction: results of a randomized, double-blind, 26-week placebo-controlled pivotal trial. *Urology* 61: 8–14
25 Atz AM, Lefler AK, Fairbrother DL, Uber WE, Bradley SM (2002) Sildenafil augments the effects of inhaled nitric oxide for postoperative pulmonary hypertensive crises. *J Thorac Cardiovasc Surg* 124: 628–629
26 Bortolotti M, Mari C, Lopilato C, Porrazzo G, Miglioli M (2000) Effects of sildenafil on esophageal motility of patients with idiopathic achalasia. *Gastroenterology* 118: 253–257
27 Ghofrani HA, Wiedemann R, Rose F, Schermuly RT, Olschewski H, Weissmann N, Gunther A, Walmrath D, Seeger W, Grimminger F (2002) Sildenafil for treatment of lung fibrosis and pulmonary hypertension: a randomized controlled trial. *Lancet* 360: 895–900
28 Jackson G, Chambers J (2002) Sildenafil for primary pulmonary hypertension: short and long term symptomatic benefit. *Int J Clin Pract* 56: 397–398
29 Lichenstein JR (2003) Use of sildenafil citrate in Raynaud's phenomenon: comment on the article by Thompson et al. *Arthritis Rheum* (Letter) 48: 282–283
30 Nakamizo T, Kawamata J, Yoshida K, Kawai Y, Kanki R, Sawada H, Kihara T, Yamashita H, Shibasaki H, Akaike A et al. (2003) Phosphodiesterase inhibitors are neuroprotective to cultured spinal motor neurons. *J Neurosci Res* 71: 485–495
31 Zhang R, Wang Y, Zhang L, Zhang Z, Tsang W, Lu M, Zhang L, Chopp M (2002) Sildenafil (Viagra) induces neurogenesis and promotes functional recovery after stroke in rats. *Stroke* 33: 2675–2680
32 Vardi Y, Klein L, Nassar S, Sprecher E, Gruenwald I (2002) Effects of sildenafil citrate (Viagra) on blood pressure in normotensive and hypertensive men. *Urology* 59: 747–752
33 Sadovsky R, Miller T, Moskowitz M, Hackett G (2001) Three year update of sildenafil citrate (Viagra) efficacy and safety. *Int J Clin Pract* 55: 115–128
34 Herrmann HC, Chang G, Klugherz BD, Mahoney PD (2000) Hemodynamic effects of sildenafil in men with severe coronary artery disease. *N Engl J Med* 342: 1622–2626
35 Padma-Nathan H, Eardley I, Kloner RA, Laties AM, Montorsi F (2003) A 4-year update on the safety of sildenafil citrate (Viagra). *Urology* 60: 67–90
36 Jackson G, Benjamin N, Jackson N, Allen MJ (1999) Effects of sildenafil citrate on human hemodynamics. *Am J Cardiol* 83: 13–20C
37 Kloner RA, Brown M, Prisant LM, Collins M (2001) Effect of sildenafil in patients with erectile dysfunction taking antihypertensive therapy. Sildenafil Study Group. *Am J Hypertens* 14: 70–73
38 Guay AT, Perez JB, Jacobson J, Newton RA (2001) Efficacy and safety of sildenafil citrate for treatment of erectile dysfunction in a population with associated organic risk factors. *J Androl* 22: 793–797
39 Conti CR, Pepine CJ, Sweeney M (1999) Efficacy and safety of sildenafil citrate in the treatment of erectile dysfunction in patients with ischemic heart disease. *Am J Cardiol* 83: 29C–34C
40 Lim PH, Moorthy P, Benton KG (2002) The clinical safety of Viagra. *Ann NY Acad Sci* 962: 378–388
41 Wallis RM, Corbin JD, Francis SH, Ellis P (1999) Tissue distribution of phosphodiesterase families and the effects of sildenafil on tissue cyclic nucleotides, platelet function, and the contractile

responses of trabeculae carneae and aortic rings *in vitro. Am J Cardiol* 83: 3C–12C
42 Gopal VG, Francis SH, Corbin JD (2001) Allosteric sites of phosphodiesterase-5 (PDE5): A potential role in negative feedback regulation of cGMP signaling in corpus cavernosum. *Eur J Biochem* 268: 3304–3312
43 Maurice DM, Haslam RJ (1990) Molecular basis of the synergistic inhibition of platelet function by nitrovasodilators and activators of adenylate cyclase: inhibition of cyclic AMP breakdown by cyclic GMP. *Mol Pharmacol* 37: 671–681
44 Ignarro LJ, Degnan JN, Baricos WH, Kadowitz PJ, Wolin MS (1982) Activation of purified guanylate cyclase by nitric oxide requires heme: comparison of heme-deficient, heme-reconstituted and heme-containing forms of soluble enzyme from bovine lung. *Biochim Biophys Acta* 718: 49–59
45 Ignarro LJ, Bush PA, Buga GM, Wood KS, Fukuto JM, Rajfer J (1990) Nitric oxide and cyclic GMP formation upon electrical field stimulation cause relaxation of corpus cavernosum smooth muscle. *Biochem Biophys Res Commun* 170: 843–850
46 Rajfer J, Aronson WJ, Bush PA, Corcy FJ, Ignarro LJ (1992) Nitric oxide as a mediator of relaxation of the corpus cavernosum in response to nonadrenergic, noncholinergic neurotransmission. *N Engl J Med* 326: 90–94
47 Webb DJ, Muirhead GJ, Wulff M, Sutton JA, Levi R, Dinsmore WW (2000) Sildenafil citrate potentiates the hypotensive effects of nitric oxide donor drugs in male patients with stable angina. *J Am Coll Cardiol* 36: 25–31
48 Turko IV, Ballard SA, Francis SH, Corbin JD (1999) Inhibition of cyclic GMP binding, cyclic GMP specific phosphodiesterase 5 by sildenafil and related compounds. *Mol Pharmacol* 56: 124–130
49 McAllister-Lucas LM, Sonnenburg WK, Kadlecek A, Seger D, Trong HL, Colbran JL, Thomas MK, Walsh KA, Francis SH et al. (1993) The structure of a bovine lung cGMP-binding cGMP-specific phosphodiesterase deduced from a cDNA clone. *J Biol Chem* 268: 22863–22873
50 Loughney K, Hill TR, Florio VA, Uher L, Rosman GJ, Wolda SL, Jones BA, Howard ML, McAllister-Lucas LM, Sonnenburg WK et al. (1998) Isolation and characterization of cDNAs encoding PDE5A, a human cGMP binding, cGMP specific 3',5'-cyclic nucleotide phosphodiesterase. *Gene* 216: 139–147
51 Kotera J, Fujishige K, Akatasuka H, Imai Y, Yanaka N Omori K (1998) Novel alternative splice variants of cGMP-binding cGMP-specific phosphodiesterase. *J Biol Chem* 273: 26982–26990
52 Lin C-S, Lau A, Tu R, Lue TF (2000) Expression of three isoforms of cGMP-binding, cGMP-specific phosphodiesterase (PDE5) in human penile tissue. *Biochem Biophys Res Commun* 268: 628–635
53 Francis SH, Thomas MK, Corbin JD (1990) Cyclic GMP-binding cyclic GMP-specific phosphodiesterase from lung. In: J Beavo, M Houslay (eds): *Cyclic Nucleotide Phosphodiesterases: Structure, Regulation and Drug Action*. John Wiley & Sons, New York, 117–140
54 Wyatt TA, Naftilan AJ, Francis SH, Corbin JD (1998) ANF elicits phosphorylation of the cGMP phosphodiesterase in vascular smooth muscle cells. *Am J Physiol* 274: H448–H455
55 Corbin JD, Turko IV, Beasley A, Francis SH (2000) Phosphorylation of phosphodiesterase-5 by cyclic nucleotide-dependent protein kinase alters its catalytic and allosteric cGMP-binding activities. *Eur J Biochem* 267: 2760–2767
56 Rybalkin SD, Rybalkina IG, Feil R, Hofmann F, Beavo JA (2002) Regulation of cGMP-specific phosphodiesterase (PDE5) phosphorylation in smooth muscle cells. *J Biol Chem* 277: 3310–3317
57 Rybalkin SD, Rybalkina IG, Shimizu-Albergine M, Tang XB, Beavo JA (2003) PDE5 is converted to an activated state upon cGMP binding to the GAF A domain. *EMBO J* 22: 469–478
58 Mullershausen F, Friebe A, Feil R, Thompson WJ, Koesling D (2003) Direct activation of PDE5 by cGMP: long-term effects within NO/cGMP signaling. *J Cell Biol* 160: 719–727
59 Corbin JD, Blount MA, Weeks JL, 2nd, Beasley A, Kuhn KP, Yew S, Ho J, Saidi LF, Hurley JH, Kotera J, Fancis SH (2003) [^{3}H]Sildenafil binding to phosphodiesterase-5 is specific, kinetically heterogeneous, and stimulated by cGMP. *Mol Pharmacol* 63: 1364–1372
60 Walker DK, Ackland MJ, James GC, Muirhead GJ, Rance DJ, Wastall P, Wright PA (1999) Pharmacokinetics and metabolism of sildenafil in mouse, rat, rabbit, dog and man. *Xenobiotica* 29: 297–310
61 Muirhead GJ, Allen MJ, James GC, Pearson J, Rance DJ, Houston AC, Dewland PM (1996) Pharmacokinetics of sildenafil (Viagra), a selective cGMP PDE5 inhibitor, after single oral doses in fasted and fed healthy volunteers. *Br J Clin Pharmacol* 42: 268P
62 Viagra labeling information (1997) FDA submission, Pfizer (New York, NY)

63 Laties AM (2001) Ocular effects of sildenafil citrate. *Sexual Dysfunction in Medicine* 2: 99–105
64 Goldstein I, Lue TF, Padma-Nathan H, Rosen RC, Steers WD, Wicker PA (1998) Oral sildenafil in the treatment of erectile dysfunction. *N Engl J Med* 338: 1397–1404
65 Nandwani R, Gourlay Y (1999) Possible interaction between sildenafil and HIV combination therapy. *Lancet* 353: 840
66 Muirhead GJ, Wulff MB, Fielding A, Kleinermans D, Buss N (2000) Pharmacokinetic interaction between sildenafil and saquinavir/ritonavir. *Br J Clin Pharmacol* 50: 99–107
67 Berkels R, Klotz T, Sticht G, Englemann U, Klaus W (2001) Modulation of human platelet aggregation by the phosphodiesterase type 5 inhibitor sildenafil. *J Cardiovasc Pharmacol* 37: 413–421
68 Halcox JP, Nour KR, Zalos G, Mincemoyer RA, Waclawiw M, Rivera CE, Willie G, Ellahham S, Quyyumi AA (2002) The effect of sildenafil on human vascular function, platelet activation, and myocardial ischemia. *J Am Coll Cardiol* 40: 1232–1240
69 Muirhead GJ, Wilner K, Colburn W, Haug-Pihale G, Rouviex B (2002) The effects of age and renal and hepatic impairment on the pharmacokinetics of sildenafil citrate. *Br J Clin Pharmacol* 53: 21S–30S
70 Hyland R, Roe EGH, Jones BC, Smith DA (2001) Identification of the cytochrome P450 enzyme involved in demethylation of sildenafil. *Br J Clin Pharmacol* 51: 239–248
71 Warrington JS, Shader RI, von Moltke LL, Greenblatt DJ (2000) *In vitro* biotransformation of sildenafil (Viagra): identification of human cytochromes and potential drug interactions. *Drug Metab Dis* 28: 392–397

Sildenafil citrate: a 5-year update on the worldwide treatment of 20 million men with erectile dysfunction

Culley C. Carson III

Division of Urology, School of Medicine, University of North Carolina at Chapel Hill, 2140 Bioinformatics Building Campus Box 7235, Chapel Hill, NC 27599-7235, USA

Introduction

Erectile dysfunction (ED) is a highly prevalent condition affecting as many as 50% of men 40 years of age or older. ED, defined as "the inability to obtain and maintain erections sufficiently to allow sexual intercourse", has been widely studied for many decades. The first breakthrough in effective, minimally invasive, safe treatment was the establishment of the oral agent sildenafil citrate for the treatment of ED. ED can be caused by systemic diseases, especially those effecting the vascular system or nervous system and can be associated with primary or secondary psychologic complaints. The incidence and prevalence of ED increases with age, with the incidence doubling with each decade of life after the age of 50. As men age, conditions of aging appear and medications for treating these chronic conditions are used widely. Many of the medications used to treat chronic conditions such as diabetes, hypercholesterolemia, renal disease, and depression will exacerbate ED. The most common culprits for medically-induced ED are antihypertensive drugs and antidepressants. Pelvic surgery and radiation therapy for conditions of aging such as prostate cancer and colon cancer may also produce ED. Life style conditions, such as cigarette smoking, obesity, sedentary life style, and the metabolic syndrome are also associated with ED.

Sildenafil

Sildenafil citrate (Viagra, Pfizer, New York) was first introduced to the US market in April 1998. During the past five years, more than 20 million men worldwide have been treated with Viagra, with more than 10 million men in the US alone. More than 24 million prescriptions have been dispensed. This accounts for more than 100 million prescriptions worldwide, including 700 million tablets. It has been estimated that 8.6 sildenafil tablets are dispensed every second worldwide. More than 500,000 physicians are prescribing sildenafil; it has been approved, widely used, and accepted in more than 110 countries. In the

US, the average use of sildenafil is once approximately every 8–11 days. If one reviews the distribution of patients taking sildenafil by age, it is apparent that the reflection of sildenafil patients in the US is very similar to those involved in the drug trials performed for sildenafil before it was approved. Thus, the data from the pivotal trials and other trials reflect the use, efficacy, and safety of sildenafil adequately and accurately. In these studies, 6% of the patients were younger than 40 years of age, 16% were between the ages of 40–49 years, 31% were between the ages of 50–59 years, 29% were between the ages of 60–69 years, 16% were between the ages of 70–79, and 2% were older than 80 years of age [1]. A study by Shakir et al. [2] confirmed that trial participants in Europe and the US mirrored this postmarketing use of sildenafil. Worldwide experience has been reported with efficacy response rates similar throughout the world. This includes countries in North America, Europe, Latin American, Africa, and Asia. Prescriptions for sildenafil are written by every type of specialist. In the US, urologists account for 15% of these prescriptions; 63% of sildenafil prescriptions are being written by general practitioners. Cardiologists account for 3%, psychiatrists 2%, endocrinologists and oncologists 1% each, and other medical specialties 15% [1].

Pharmacology

Sildenafil is the first of the phosphodiesterase type-5 (PDE-5) agents clinically available for the treatment of men with ED of all ages, severities and etiologies. These agents, which facilitate erectile function, require sexual stimulation for their activity. Sildenafil works through the secondary neurotransmitter system of nitric oxide (NO). When stimulation of the corpora cavernosa occurs from central nervous system (CNS) activation through physical, psychologic, or other CNS stimuli, NO enters the cell through the cell membrane and activates the guanylate cyclase system to increase the production of cyclic guanylate monophosphate (cGMP) [3]. This agent, which produces smooth muscle relaxation in the corpus cavernosum by activating the G protein and producing a decrease in intracellular calcium and subsequent smooth muscle relaxation, facilitates erectile function by this smooth muscle relaxation. NO is produced through both nerve endings and from endothelial cells. NO is manufactured in each of these locations by the conversion of L-Arginine to nitric oxide through the enzyme nitric oxide synthase. cGMP is broken down in the smooth muscle cell by the enzyme phosphodiesterase (PDE), with PDE-5 being the most prevalent in the corpus cavernosum smooth muscle tissue. By inhibiting this enzyme, sildenafil facilitates erectile function by increasing the intramuscular concentration of cGMP. With higher levels of cGMP, erections are more likely to occur in patients who are deficient in erectile function and the duration of erectile function also will increase in those patients for whom decreased erectile duration is a problem. Thus, sildenafil facilitates erection through normal physiologic processes and does not produce erection as other agents do (e.g., alprostadil).

Clinical trial outcome measurement

The clinical efficacy of erectogenic agents can be measured with a variety of instruments. The most common instrument is the global efficacy assessment question (GAQ), which asks patients "Did the treatment that you used for your erections improve your erections during the time of therapy". Although this is an important endpoint measure, it is not as rigorous as some other measures. However, it is the standard measurement for most clinical studies of erectile function agents. Other endpoints include a measurement of the percent of successful attempts at intercourse and a measurement of the International Index of Erectile Function (IIEF) with changes in the erectile function domain score from baseline to study completion. Using this score, values less than 10 suggest severe ED and values higher than 26 suggest normal erectile function. Other methods used for measurement include sexual encounter profile questions (SEP). The SEP-2 asks, "Were you able to insert your penis into your partner's vagina" and the SEP-3 asks, "Did your erections last long enough to have successful intercourse?"

Clinical trials

Sildenafil citrate has been evaluated clinically and experimentally with each of these endpoint measurements. Initial studies using the GAQ in more than 3000 patients with a variety of etiologies of ED in 11 double-blind, placebo-controlled, flexible-dose studies at week 12 demonstrated that placebo improved erections by 22% compared with sildenafil at 76% [4]. When querying patients about successful attempts at intercourse, this same group of patients answered 26% of the questions positively with placebo compared with 66% with sildenafil. In reviewing four randomized, double-blind, placebo-controlled, fixed-dose studies of 12–24 weeks duration, a placebo response was demonstrated in the group of patients queried about improvement in their erections with GAQ [4]. In this group, 24% had positive placebo response, 63% responded positively to 25 mg of sildenafil, 74% to 50 mg, and 82% to 100 mg. When observing patients of various ages, similar efficacies were demonstrated. In these studies, 24% of patients who were younger than 60 years of age responded to placebo compared with 78% of those of the same age in the sildenafil group ($P < 0.0001$); 18% of patients responded to placebo compared with 69% to sildenafil ($P < 0.0001$). In examining the severity of ED, those patients with mild to moderate ED, which is defined by IIEF erectile function domain scores of 17–21, 35% responded to placebo with improved erections compared with 87% with sildenafil ($P < 0.0001$). For patients with severe ED, which is defined by IIEF erectile function domain scores of 6–10, 12% responded to placebo compared with 65% to sildenafil ($P < 0.0001$) [4].

Sildenafil also has been effective across multiple etiologies. In the same 11 double-blind, placebo-controlled, flexible-dose studies at week 12, patients were divided into organic, psychogenic, and mixed etiologies of ED and were asked about improved erectile function at the conclusion of the study [4]. In those patients with organic ED, 18% of placebo patients responded compared with 69% of those administered sildenafil; in psychogenic patients, the response was 38% for placebo *versus* 86% for sildenafil; in mixed etiology patients, the response was 28% placebo compared with 85% for sildenafil. All of the responses were highly statistically significant at P < 0.0001 [4].

When counseling patients, it is important to identify the response of any ED treatment agent by organic etiology such that they can be counseled based on the evaluation of their risks and their likelihood to respond to treatment. Sildenafil citrate improved erectile function across etiologies [4]. In a group of patients in 11 double-blind, placebo-controlled, flexible-dose clinical trials at week 12, diabetic patients responded quite well [4]. There was an 18% placebo response for Type I diabetics *versus* 59% with sildenafil. For Type II diabetics, there was a 17% placebo response compared with a 63% sildenafil citrate response. Both of these were high statistically significant. Stuckey et al. [5] evaluated 188 Type I diabetic men with ED treated with sildenafil. In assessing SEP-4, 68.4% of the patients treated with sildenafil compared with 26.5% of those treated with placebo responded "yes" (P < 0.0001). Other endpoints also showed a high efficacy and adverse events were rare [5]. There is some reduction in response for patients with complications of diabetes. In these patients, the more complications of diabetes there are and the more difficult the diabetes is to control, the weaker the response is to sildenafil. In these studies, of those patients without any complications of diabetes, 8% of placebo patients responded compared with 69% of those administered sildenafil. With one complication of diabetes, placebo response was 12% compared with 43% for those who administered sildenafil. Of those patients with two or more complications of diabetes, 10% of those administered placebo responded compared with only 43% of the sildenafil patients [4]. Complications included neuropathy, ocular changes, and other vascular disease.

Hypertension is a well-known risk factor for ED, increasing the prevalence of ED beyond normal age increases. Although sildenafil has a mild hypotensive effect, it has been proven to be safe and effective in treated and untreated hypertensive men [6–8]. Sildenafil was also demonstrated to cause an excellent response in patients with treated and untreated hypertension who had ED [4, 8]. In a mixed group of hypertensive patients, 18% responded to placebo *versus* 70% to sildenafil (P < 0.0001). In patients taking two or more antihypertensive medications, patients were well treated and responded well to sildenafil treatment [8]. In evaluating the improvement in erections, 17.6% of patients improved with placebo compared with 71% with sildenafil. When evaluating the percent of successful attempts at intercourse, 26.1% of patients had successful intercourse on placebo compared with 62.9% on sildenafil. This translated into an increase in successful numbers of intercourse attempts per week from 0.66 for placebo to 1.6 for patients treated with sildenafil. These efficacy measures were consistent with no increase in adverse events including headache, vasodilation, or dizziness [8].

Smoking has been widely demonstrated as a risk factor for ED [9]. Furthermore, sildenafil has been shown to augment endothelial-dependent vasodilation in smokers [10]. Sildenafil citrate appears to be effective in improving erections in patients who are current and ex-smokers. In those patients who smoke currently, 25% responded to placebo compared with 80% to sildenafil. In ex-smokers, there was an 18% placebo response compared with a 76% sildenafil response. When comparing these groups with patients who had never smoked, 26% of patients responded to placebo and 74% responded to sildenafil citrate. All of these values had a statistical significance of $P < 0.0001$ [4].

Psychologic etiologies are also significant risk factors for ED. Depression has been demonstrated in a number of epidemiologic studies to be as high as second behind severe vascular and cardiac disease as a cause of ED [11, 12]. It is also widely known that patients who are more severely disabled by depression are more severely disabled by their ED [13]. In a 12-week study of placebo *versus* sildenafil that reviewed the response of patients with mild untreated depression and ED, patients who responded to sildenafil or placebo had a decrease in their depression scores, which was measured by the Beck's depression inventory and the Hamilton depression rating scale [14]. Patients taking selective serotonin reuptake inhibitors (SSRIs) for depression also respond well to sildenafil treatment. Nurnberg et al. [13] compared more than 3000 patients with significant depression, more than 1000 of them taking placebo and more than 2000 of them taking sildenafil. The endpoint measured was the ability to obtain and maintain an erection satisfactory for vaginal penetration comparing placebo, baseline, and treatment at the end of 12 weeks. In all of the groups, sildenafil improved erectile function more than placebo and baseline at a highly statistically significant value. Although the patients treated with SSRIs had a slightly less response than those without SSRI treatment, the difference between SSRI-treated and untreated patients was not statistically significant. In patients with major depressive disorder (MDD) treated for depression, sildenafil appears to be effective in treating pre-existing ED that persists after MDD therapy [15]. In SSRI-involved ED, sildenafil has been demonstrated to be effective in open-label and placebo-controlled trials [13, 16].

Other neurologic conditions can also be treated effectively with sildenafil. In a study by Giuliano et al. [17], a group of patients with spinal cord injury were queried regarding their ability to maintain an erection with baseline placebo compared with sildenafil. The responses to sildenafil, although less than those of patients without neurologic injury, were highly statistically significantly better than those of baseline or placebo ($p < 0.0001$) [17]. In men with spinal cord injury, ED treatment will improve erections and have a major impact on self-esteem and quality of life [18]. Similarly, in a group of patients with multiple sclerosis who were administered the global efficacy question,

Fausier et al. identified a placebo response of 24% compared with an 89% sildenafil response [19].

Other neurologic conditions and injuries will also lead to significant ED. Pelvic surgery, including colectomy and radical prostatectomy, may produce ED [20]. It is well known that patients undergoing radical prostatectomy, whether it is nerve-sparing or not, will have a risk of developing ED [21]. The older the patient undergoing radical prostatectomy, the more likely ED will ensue [22]. It has been demonstrated that erectile function improves with time from surgery if careful nerve-sparing operations have been performed [23]. Erectile function appears to improve for as long as four years after surgery, supporting the repeated trials of medication for post-operative ED even if early trials failed [23]. In a single institution study, Raina et al. [24] reported a comparison of 53 bilateral nerve-sparing, 12 unilateral nerve-sparing, and 26 non-nerve sparing radical prostatectomy patients measuring their IIEF response, ability to penetrate, and spousal satisfaction with sildenafil after undergoing a radical prostatectomy. The bilateral nerve-sparing patients reported the ability to penetrate 71.7% of the time compared with 50% for unilateral nerve-sparing patients and 15% non-nerve sparing patients ($P < 0.0001$). Further confirmation of these data were from a spousal satisfaction questionnaire on which 66% of the spouse of bilateral nerve-sparing patients reported satisfaction compared with 41.6% unilateral nerve-sparing patients and 15.4% non-nerve sparing patients ($P < 0.0001$). In 1999, Hong et al. [25] reported on 316 patients who underwent radical prostatectomies 1–4 years before the study. 119 of 316 patients were treated with sildenafil and evaluated with the Erectile Dysfunction Inventory of Treatment Satisfaction questionnaire. Pre- and post-operative sexual function was evaluated with the O'Leary Brief Sexual Function Inventory and a pre- and post-operative sexual history. Of these patients, 92% had erections sufficient for vaginal penetration pre-operatively and 95% of all of the patients in the study had bilateral nerve-sparing radical prostatectomies. In their study, the authors demonstrated a gradual increase in the response to sildenafil over time after surgery. Of those patients 0–6 months after surgery, 26% responded compared with 36% at 6–12 months, 50% at 12–18 months, and 60% at 18–24 months. These data have been confirmed in other studies with and without sildenafil and have demonstrated the gradual improvement, healing, and resolution of neuropathy for as long as 3–4 years after undergoing a radical prostatectomy [4, 24]. The controversy regarding the use of sildenafil and other agents for prophylaxis after radical prostatectomy continues. Although the treatment of ED after a radical retropubic prostatectomy for carcinoma of the prostate is well established and sildenafil appears to be effective, the question of the use of pharmacologic agents for prophylaxis during the healing period after a radical prostatectomy continues to be controversial. The possibility of prophylaxis in these patients was first described by Montorsi et al. [26]. These investigators compared two groups of men after they underwent a nerve-sparing radical prostatectomy. The first group was treated with alprostadil injections three times weekly for 12 weeks beginning four weeks after radical prostatectomy; the second group of men were not treated. Patients were assessed at the six-month follow-up with a history, physical examination, and colored Doppler stenography studies. The group of patients treated with regular alprostadil injections had an 80% likelihood of recovery of erectile function compared with 20% of those patients without treatment [26]. Padma-Nathan et al. [27] described the use of sildenafil citrate prophylaxis in the post-radical prostatectomy patients. After a nerve-sparing radical prostatectomy, 76 men with normal pre-operative erectile function were administered placebo or 50 or 100 mg of sildenafil for 36 weeks. In this double-blind study, erectile function was assessed eight weeks after the termination of the study using the GAQ, IIEF, and nocturnal penile tumescence assessments. Rigorous endpoints included a combined score of higher than eight for IIEF question three and four and a positive GAQ. Of those patients treated with sildenafil, 27% experienced a return of spontaneous erections within 48 weeks of surgery compared with 4% in the placebo group. Although these data are preliminary, it is clear that nightly administration of sildenafil after a radical prostatectomy holds promise for improving the return of spontaneous erections in men undergoing bilateral nerve-sparing prostatectomy [27, 28].

Sildenafil in the cardiac patient

The concern regarding the cardiac effects of sildenafil and other PDE-5 inhibitors has been widely reported in scientific and lay press [1, 29]. Studies investigating the effectiveness of sildenafil on those patients with ED associated with ischemic heart disease has been reported. In a group of more than 300 patients with ischemic heart disease treated with placebo or sildenafil, 16% of the patients administered placebo reported improved erections compared with 62% of the patients taking sildenafil [29]. A number of studies have also been carried out to evaluate the effect of sildenafil and other PDE-5 inhibitors on blood pressure, cardiac function, and QRS complex prolongation. Blood pressure studies by Zusman et al. [30] investigated the comparison of blood pressure changes in a group of patients taking no associated medication and a variety of antihypertensives. Although there were some decreases in sildenafil-treated patients in each group, the numbers rarely were statistically significant. The agents evaluated included diuretics, α-blockers, β-blockers, angiotensin-converting enzyme inhibitors, and calcium channel blocking medications. The greatest blood pressure decline after the administration of sildenafil was identified in the patients taking no antihypertensive medication and the calcium channel blocker-treated patients [30, 31]. Hermann et al. [31] investigated the effect of sildenafil treatment on a group of carefully followed patients in a cardiac catheterization laboratory. In their study there was a demonstration of no increase in angina, timed angina, or deterioration of cardiac function, but rather some facilitation of flow-mediated coronary artery dilation compared with those patients treated with placebo. In investigating the cardiac events and mortality associated with patients with ED treated with sildenafil, a compilation of studies and postmarketing experience has demonstrated no difference in the incidence of cardiac abnormality or death [32, 33]. In clinical trials, placebo- and sildenafil-treated patients had a cardiovascular system mortality report in 0.3 per 100 patient years, with confidence interval equivalent in both groups [30]. Death from myocardial infarction (MI) and ischemic heart disease after the trial was similar, measuring 0.26 per 100 patient years [32]. This compares favorably with the expected rate of cardiac abnormalities in this age group of patients that measure 0.6 per 100 patient years.

Adverse events are a concern for patients taking any medication. PDE-5 inhibitors and sildenafil citrate in particular exhibit mild, but ever-present adverse events [1, 4]. The adverse events most commonly reported include headache (approximately 7%), vasodilation symptoms (7%), dizziness (2%), dyspepsia (1.8%), rhinitis (1.4%), and abnormal vision usually characterized as blue tint to the vision (approximately 1.2%) [1, 32]. Discontinuation of the use of medication and incidence of adverse events essentially was equivalent in placebo- and sildenafil-treated patients in the pivotal trials [30]. These adverse events accounted for approximately 2% of patients discontinuing trials. In a study by Carson [34], adverse events were investigated from the initiation of treatment in pivotal trials through 16 weeks after the beginning of the trials. In this study, all of the adverse events declined precipitously from the initiation of treatment to the ultimate conclusion of the study. All of the adverse events, except for abnormal vision and dyspepsia, declined dramatically. Headache declined from almost 7% to less than 1% during these 16 weeks as did vasodilation, which declined from 7% to less than 1%; dizziness declined to less than 1% and rhinitis from 1.4% to less than 0.5%. Thus, the incidents of adverse events with sildenafil treatment appeared to decline over time; this despite the fact that as many as two-thirds of the patients in this study increased dosage during the trial. With ongoing treatment and at the conclusion of the study, adverse event frequency with sildenafil and placebo treatment essentially was identical.

Non-responders and incomplete responses are troublesome for all medical treatment. Sildenafil is no different in this inadequate response. Indeed, as many as 20% of patients will respond poorly to sildenafil and discontinue treatment as a result of poor response. Most common causes of inadequate response are administration problems [28]. Patients can optimize the response to sildenafil by taking the medication on an empty stomach, waiting 30–60 min before the initiation of sexual function, and, most importantly, optimizing sexual stimulation for increased and improved response. Similarly, patients should escalate their dosage until they reach 100 mg or until the response is appropriate. Dose titration is especially important because most of the patients treated with sildenafil require 100 mg. McCullough et al. [28] have also demonstrated that multiple attempts at sildenafil use are required before an optimal response can be recognized. Their study demonstrated that 6–8 trials of Viagra at optimal doses were necessary to produce satisfactory responses in most patients. Therefore, it is important for the physician to diagnose ED and prescribe appropriate medication, but also to coach the patients regarding appropriate use, titration, and stimulation associated with success [35]. Overall, the most important concept is the improvement in quality of life for patients and their partners. In 2000, McMahon et al. [36] reported a quality- of-life questionnaire administered to 140 patients and 107 partners at baseline and after sildenafil treatment. They demonstrated a highly statistically significant improvement in quality of life for the patients and their partners once ED was managed and sexual function returned to a satisfactory level. In open-label extension studies, a mirror of community results, Montorsi et al. [37] reported on 2618 men treated for up to three years. 73% of these patients required 100 mg of Viagra and 99% reported an improvement in sexual ability, with 96% satisfied with their treatment. Only 1.6% of these responders ultimately discontinued medication as a result of lack of efficacy. In a study reported by Padma-Nathan et al. [27], there was as much as a three-fold increase in the resolution and recovery of spontaneous erections when using sildenafil prophylactically after a bilateral nerve-sparing radical prostatectomy. These patients, treated four weeks after the radical prostatectomy and continuing for 12 weeks, had substantial improvement with little morbidity and few adverse events.

Sildenafil citrate appears to be effective for long-term use. A recent report by El-Galley et al. [38] suggested a possible tachyphylaxis when sildenafil citrate is taken for long periods of time. The authors reviewed a telephone survey of 151 men treated for one year with sildenafil. A follow-up telephone questionnaire was carried out two years after the initiation of therapy. The response rate of this questionnaire was 54% (82 of 151). Of these long-term responders, 41 of 69 (59%) were still using sildenafil with satisfactory response. 37% of patients required dose escalation during treatment. 17% discontinued use because of loss of efficacy. Although this study describes a tachyphylaxis effect, a small number of responders to the telephone questionnaire and the lack of exclusion of underlying disease progression, concomitant therapy initiation, and other confounding factors produce a somewhat confusing picture from this effect. Laboratory studies and other large long-term studies have failed to demonstrate tachyphylaxis affect [39].

With its serum half life (T1/2) of four hours, sildenafil lasts from 4–8 hours after ingestion. The maximum serum concentration (Tmax) is achieved at approximately one hour and it was previously thought that the onset was only 60 min after ingestion. Recent studies, however, have shown that onset is as early as 14 min in more than one-third of men and less than 20 min in more than half the men tested with a journal and stop watch study. Thus, many men may be able to have adequate intercourse within half an hour of medication administration [40].

Sildenafil for non-erectile dysfunction indications

Multiple studies have been carried out to investigate the use of sildenafil for indications other than ED. Sildenafil was designed initially as a coronary artery dilating agent [40]. Its effectiveness through endothelial cell function suggests other possible uses for this agent [40]. The correlation of ED with benign prostatic hyperplasia and lower urinary tract symptoms has been demonstrated epidemiologically [41]. The treatment of each of these conditions independently appears to have an influence on the other condition. Sairam et al. [42] investigated the effect of sildenafil on lower urinary tract symptoms. Questionnaire investigation using the IIEF and the International Prostate Symptoms System (IPSS) was carried out at baseline and at each return after the initiation of oral sildenafil treatment. Patients were followed at one and three months after the initiation of treatment. During treatment with sildenafil, IPSS scores improved and ED was satisfactorily managed in these men with lower urinary tract symptoms [42]. In other non-ED effects, PDE-5 inhibitors have been demonstrated to improve refractory, primary, and secondary pulmonary hypertension. Sildenafil has been used successfully as a single agent and as an adjunct to inhaled iloprost in improving functional status for patients who are difficult to treat [43]. Similarly, pulmonary artery function has been improved in patients with severe congestive heart failure [44]. Thus, sildenafil improves erectile function in patients with cardiac disease and congestive heart failure and may have some impact on difficult-to-treat parameters of the underlying cardiac disease [45–48].

An important area of research and investigation for the use of PDE-inhibiting agents for sexual dysfunction is in female sexual dysfunction. Although the physiologic principles of female dysfunction differ substantially from those of ED, many agents effective for male sexual dysfunction may be applicable for female sexual dysfunction. Difficulties with classification, diagnosis, and appropriate choice of agents continue to be significant in female sexual dysfunction [49]. Modelska and Cummings [50] have reviewed the randomized, placebo-controlled clinical trial for female sexual dysfunction in postmenopausal women. They concluded that adequate evidence is not available to clearly demonstrate adequate response to sildenafil for widespread use for female sexual dysfunction [51]. Basson et al. [52] performed a placebo-controlled trial using global efficacy, life satisfaction checklist, and sexual function questionnaires and adverse event recordings to evaluate 577 well-estrogenized and 204 estrogen-deficient women with female sexual dysfunction. These patients were diagnosed with female sexual arousal disorder. Efficacy between placebo and sildenafil were not significant for the patients or their partners. Their conclusion was that the physiologic effect of sildenafil was not adequate to perceive improvement in sexual response in estrogenized or estrogen-deficient women with female sexual arousal disorder.

Conclusions

The four years of sildenafil availability throughout the world has changed the complexion of ED treatment for patients and their partners. Sildenafil is effective in 60–80% of patients, regardless of the duration of their ED, etiology of ED, severity of ED, or age. With few adverse events, sildenafil can be prescribed safely and reliably for these patients with a high degree of patient and partner satisfaction [1]. At one year of sildenafil treatment, more than 2400 men were queried about their satisfaction with treatment; 96% were satisfied with the treatment effect on erections and 99% reported an improved ability to engage in sexual activity [4]. Three years after beginning treatment, the data were durable, with 95.1% and 99.8% responses, respectively [4]. Adverse events are few and the effects on cardiac function and interaction with other medications are minimal. It is clear that the past five years have been a revolutionary time in the treatment of ED. Viagra, with its large quantity of data in large numbers of patients who have been followed for many years and with subpopulations of all etiologies, has demonstrated the international impact sildenafil citrate has had on patients with ED.

References

1 Carson CC (2003) Long-term use of sildenafil. *Expert Opin Pharmacother* 4: 397–405
2 Shakir SA, Wilton LV, Boshier A, Layton D, Heeley E (2001) Cardiovascular events in users of Sildenafil: results first phase of prescription event monitoring in England. *BMJ* 322: 651–652
3 Corbin JD, Francis SH, Webb DJ (2002) Phosphodiesterase type 5 as a pharmacologic target in erectile dysfunction. *Urology* 60: 4–11
4 Carson CC, Burnett AL, Levine LA, Nehra A (2002) The efficacy of Sildenafil citrate (Viagra) in patients with erectile dysfunction. *Urology* 60: 12–27
5 Stuckey BG, Jadzinsky MN, Murphy LJ, Montorsi F, Kadioglu A, Fraige F, Manzano P, Deerochanawong C (2003) Sildenafil citrate for the treatment of erectile dysfunction in men with type I diabetes mellitus a controlled trial. *Diabetes Care* 26: 279–284
6 Vardi Y, Klein L, Nassar S, Sprecher E, Gruenwald I (2002) Effects of Sildenafil citrate on blood pressure in normotensive and hypertensive men. *Urology* 59: 747–752
7 Jackson G, Benjamin N, Jackson N, Allen MJ (1999) Effects of sildenafil citrate on human hemodynamics. *Am J Cardiol* 83 (Suppl): 13C–20C
8 Kloner RA, Brown M, Prisant LM, Collins M (2001) Effects of Sildenafil in patients with erectile dysfunction taking antihypertensive therapy. Sildenafil Study Group. *Am J Hypertens* 14: 70–73
9 McVary KT, Carrier S, Wessells H (2001) Subcommittee on Smoking and Erectile Dysfunction Socioeconomic Committee, Sexual Medicine Society of North America: Smoking and erectile dysfunction: evidence-based analysis. *J Urol* 166: 1624–1632
10 Kimura M, Higashi Y, Hara K, Noma K, Sasaki S, Nakagawa K, Goto C, Oshima T, Yoshizumi M, Chayama K (2003) PDE5 inhibitor Sildenafil citrate augments endothelin dependent vasodilation in smokers. *Hypertension* 41: 1106–1110
11 Feldman HA, Goldstein I, Hatzichristou DG, Krane RJ, McKinlay JB (1994) Impotence and its medical and psychologic correlates: results of the Massachusetts Male Aging Study. *J Urol* 151: 54–61
12 Benet AE, Melman A (1995) The epidemiology of erectile dysfunction. *Urol Clin North Am* 22: 699–709
13 Nurnberg HG, Seidman SN, Gelenberg AJ, Fava M, Rosen R, Shabsigh R (2002) Depression, antidepressant therapies, and erectile dysfunction: clinical trials of Sildenafil citrate (Viagra) in treated and untreated patients with depression. *Urology* 60: 58–66
14 Seidman SN, Roose SP, Menza MA, Shabsigh R, Rosen RC (2001) Treatment of erectile dysfunction in men with depressive symptoms: results of a placebo controlled trial with Sildenafil citrate. *Am J Psychiatry* 158: 1623–1630
15 Tignol JL, Benkert O (2001) Sildenafil citrate effectively treats erectile dysfunction in men who have been successfully treated for depression. Presented at the American Psychiatric Association

Meeting, New Orleans, 10 May 2001
16 Fava M, Rankin MA, Alpert JE (1998) An open trial of oral Sildenafil in antidepressant induced sexual dysfunction. *Psychother Psychosom* 67: 328–331
17 Giuliano F, Hultling C, El Masry WS, Smith MD, Osterloh IH, Orr M, Maytom M (1999) Randomized trial of Sildenafil for the treatment of erectile dysfunction in spinal cord injury. *Ann Neurol* 46: 15–21
18 Hultling C, Giuliano F, Quirk F, Pena B, Mishra A, Smith MD (2000) Quality of life in spinal cord injury patients receiving Viagra (Sildenafil citrate) for the treatment of erectile dysfunction. *Spinal Cord* 38: 363–370
19 Fausler C, Miller J, Shariel M (1999) Viagra (Sildenafil citrate) for the treatment of men with multiple sclerosis. *Neurol* 46: 497
20 Dozois RR, Nelson H, Metcalf AM (1993) Sexual function after ileo anal anastomoses. *Ann Chir* 47: 1009–1013
21 Mueleman EJ, Mulders PF (2003) Erectile function after radical prostatectomy: a review. *Eur Urol* 43: 95–101
22 Moul JW, Mooneyhan RM, Kao TC, McLeod DG, Cruess DF (1998) Preoperative and operative factors predict incontinence, impotence, and stricture after radical prostatectomy. *Prostate Cancer Prostatic Dis* 1: 242–249
23 Begg CB, Riedel ER, Bach PB, Kattan MW, Schrag D, Warren JL, Scardino PT (2002) Variations in morbidity after radical prostatectomy. *N Engl J Med* 346: 1138–1144
24 Raina R, Lakin MM, Agarwal A, Sharma R, Goyal KK, Montague DK, Klein E, Zippe CD (2003) Long-term effect of Sildenafil citrate on erectile dysfunction after radical prostatectomy: a 3-year follow-up. *Urology* 62: 110–115
25 Hong EK, Lepor H, McCullough AR (1999) Time dependent patient satisfaction with Sildenafil for erectile dysfunction (ED) after nerve sparing radical retropubic prostatectomy (RRP). *Int J Impot Res* 11 (Suppl 1): S15–S22
26 Montorsi F, Guazzoni G, Strambi LF, Da Pozzo LF, Nava L, Barbieri L, Rigatti P, Pizzini G, Miani A (1999) Recovery of spontaneous erectile function after nerve sparing radical retropubic prostatectomy with and without early intracavernous injection of Alprostadil: results of a prospective randomized trial. *J Urol* 161: 1914–1915
27 Padma-Nathan H, McCullough AR, Guiliani F et al. (2003) Postoperative nightly administration of Sildenafil citrate significantly improves the return of normal spontaneous erectile function after bilateral nerve sparing prostatectomy. *J Urol* 169 (Suppl): 375
28 McCullough AR, Barada JH, Fawzy A, Guay AT, Hatzichristou D (2002) Achieving treatment optimization with Sildenafil citrate in patients with erectile function. *Urology* 60: 28–37
29 Conti CR, Pepine CJ, Sweeney M (1999) Efficacy and safety of Sildenafil citrate in the treatment of erectile dysfunction in patients with ischemic heart disease. *Am J Cardiol* 83 (Suppl): 29C–34C
30 Zusman RM, Prisant LM, Brown MJ (2000) Effect of Sildenafil citrate on blood pressure and heart rate in men with erectile dysfunction taking concomitant antihypertensive medication. *J Hypertens* 18: 1865–1869
31 Herrmann HC, Chang G, Klugherz BD, Mahoney PD (2000) Hemodynamic effects of sildenafil in men with severe coronary artery disease. *N Engl J Med* 342: 1622–1626
32 Padma-Nathan H, Eardley I, Kloner RA, Laties AM, Montorsi F (2002) A 4-year update on the safety of Sildenafil citrate (Viagra). *Urology* 60: 67
33 Kloner RA (2000) Cardiovascular risk and Sildenafil. *Am J Cardiol* 86: 57F–61F
34 Carson CC (2002) Sildenafil citrate treatment for erectile dysfunction: rate of adverse events decreases with time. *J Urol* 167 (Suppl): 179
35 Barada J (2001) Successful salvage of Sildenafil (Viagra) failures: benefits of patient education and rechallenge with Sildenafil. *Int J Impot Res* 13 (Suppl 4): S49
36 McMahon CG, Samoli R, Johnson H (2000) Efficacy safety and patient acceptance of Sildenafil citrate as treatment of erectile dysfunction. *J Urol* 164: 1192–1196
37 Montorsi F, McDermott TE, Morgan R, Olsson A, Schultz A, Kirkeby HJ, Osterloh IH (1999) Efficacy and safety of fixed dose oral Sildenafil in the treatment of erectile dysfunction of various etiologies. *Urology* 53: 1011–1018
38 El-Galley R, Rutland H, Talic R, Keane T, Clark H (2001) Long-term efficacy of Sildenafil and tachyphylaxis effect. *J Urol* 166: 927–931
39 Steers WD (2002) Tachyphylaxis and phosphodiesterase type 5 inhibitors. *J Urol* 168: 207
40 Padma-Nathan H, Stecher VJ, Sweeney M, Orazem J, Tseng LJ, Deriesthal H (2003) Minimal

time to successful intercourse after sildenafil citrate: results of a randomized, double-blind, placebo-controlled trial. *Urology* 62: 400–403

41 O'Leary MP (2000) LUTS, ED, QOL: alphabet soup or real concerns in aging men? *Urology* 56 (Suppl): 7–11

42 Sairam K, Kulinskaya E, McNicholas TA, Boustead GB, Hanbury DC (2002) Sildenafil influences lower urinary tract symptoms. *BJU Int* 90: 836–891

43 Ghofrani HA, Rose F, Schermuly RT, Olschewski H, Wiedemann R, Kreckel A, Weissmann N, Ghofrani S, Enke B, Seeger W et al. (2003) Oral Sildenafil as long-term adjunct therapy to inhaled iloprost in severe pulmonary arterial hypertension. *J Am Coll Cardiol* 42: 158–164

44 Alaeddini J, Uber P, Park MH, Scott RL, Mehra MR (2003) Sildenafil and assessment of pulmonary arterial reactivity in heart failure. *Congest Heart Fail* 9: 176–178

45 Sebkhi A, Strange JW, Phillips SC, Wharton J, Wilkins MR (2003) Phosphodiesterase type 5 as a target for the treatment of hypoxia-induced pulmonary hypertension. *Circulation* 107: 3230–3235

46 Bharani A, Mathew V, Sahu A, Lunia B (2003) The efficacy and tolerability of Sildenafil in patients with moderate-to-severe pulmonary hypertension. *Indian Heart J* 55: 55–59

47 Cubillos-Garzon LA, Casas JP, Morillo CA (2003) Sildenafil in secondary pulmonary hypertension. *Int J Cardiol* 89: 101–102

48 Laupland KB, Helmersen D, Zygun DA, Viner SM (2003) Sildenafil treatment of primary pulmonary hypertension. *Can Respir J* 1: 48–50

49 Rosen RC, McKenna KE (2002) PDE-5 inhibition and sexual response: pharmacological mechanisms and clinical outcomes. *Annu Rev Sex Res* 13: 36–88

50 Modelska K, Cummings S (2003) Female sexual dysfunction in postmenopausal women: systematic review of placebo-controlled trials. *Am J Obstet Gynecol* 188: 286–293

51 Rosen RC (2002) Sexual function assessment and the role of vasoactive drugs in female sexual dysfunction. *Arch Sex Behav* 31: 439–443

52 Basson R, McInnes R, Smith MD, Hodgson G, Koppiker N (2002) Efficacy and safety of Sildenafil citrate in women with sexual dysfunction associated with female sexual arousal disorder. *J Womens Health Gend Based Med* 11: 367–377

Current and future strategies for preventing and managing erectile dysfunction following radical prostatectomy

Francesco Montorsi[1], Alberto Briganti[1], Andrea Salonia[1], Patrizio Rigatti[1] and Arthur L. Burnett[2]

[1] *Department of Urology, Universita' Vita-Salute San Raffaele, Via Olgettina 60, 20132 Milan, Italy*
[2] *Department of Urology, The Johns Hopkins Hospital, Baltimore, Maryland 21287, USA*

Introduction

Radical prostatectomy is an increasingly used therapeutic option for patients with clinically localized prostate cancer and a life expectancy of at least 10 years [1]. The pioneering work by Walsh and Donker [2] significantly contributed to the understanding of the surgical anatomy of the prostate and posed the bases for the subsequent development of the anatomic radical prostatectomy technique, i.e., a surgical approach aimed at completely excising the prostate providing optimal cancer control while maintaining the integrity of the anatomical structures devoted to the functions of urinary continence and sexual potency [3–5]. Since the initial reports on this technique an increasing number of studies have reported very satisfactory postoperative rates of urinary continence while the preservation of erectile function after surgery clearly showed to be a major challenge for most urologists [6–8]. This finding contributed to the development of an increasing interest in the elucidation of the pathophysiology of postoperative erectile dysfunction (ED) and on its potential prophylaxis and treatment [9–10]. The aim of this review article was to extensively evaluate the information coming from peer reviewed abstracts and full papers on the topic of ED and radical prostatectomy in the attempt to summarize the current knowledge in the field and elucidate future therapeutic strategies.

Materials and methods

A systematic review of the literature using Medline and CANCERLIT from January 1997 to June 2003 was conducted. Electronic searches were limited to the English language using the keywords prostatectomy, prostate cancer, erectile dysfunction and impotence. In addition, abstracts published in the journals

European Urology, The Journal of Urology and *International Journal of Impotence Research* as official proceedings of internationally known scientific societies held in the same time period were also assessed. Overall, 78 papers and abstracts related to the topic of erectile dysfunction and radical prostatectomy were identified and peer reviewed by the authors and constituted the basis for this review. Finally unpublished information known by the authors which was considered of interest for the readers was also included.

Results

Patient selection

The anatomical nerve sparing approach to the prostate is considered for patients with clinically localized prostate cancer. Clinical stages T1 and T2 patients as defined following digital rectal examination and/or trans-rectal ultrasonography of the prostate are the best candidates. As clinical stage is often correlated to both PSA values and the biopsy Gleason sum it has been suggested that a nerve sparing approach may be considered in patients with prostate specific antigen (PSA) less than 10 ng/ml and a Gleason sum ≤7 [11].

Patients being considered for a nerve sparing radical prostatectomy should be potent prior to the procedure. This is of major importance as patients who report some degree of erectile dysfunction or patients who use phosphodiesterase type 5 inhibitors (PDE-5-I) prior to the procedure are more likely to develop severe erectile dysfunction after the procedure itself [12]. The use of validated questionnaires such as the International Index of Erectile Function (IIEF) [13] may facilitate the diagnosis of preoperative erectile dysfunction during the initial patient assessment. This questionnaire assesses various domains of male sexuality including erectile and ejaculatory function, orgasm, desire and intercourse satisfaction and in the experience of the authors of this review by reviewing the results of this patient self assessment interesting baseline information are always obtained. Morphology of the corpora cavernosa deteriorates with aging and this may be correlated with the high prevalence of erectile dysfunction seen in the aging men [14]. Rates of recovery of erectile function after a nerve sparing radical prostatectomy are inversely correlated with the patient's age, i.e., best post-operative potency rates are obtained in the younger patient population and it seems reasonable to consider patients ≤65 years of age as candidates for a nerve sparing procedure [3]. Co-morbid conditions seem also to affect the recovery of spontaneous erections postoperatively as they may impact on the baseline penile haemodynamics. Namely, a concomitant diagnosis of diabetes mellitus, hypertension, ischemic heart disease, hypercholesterolemia or history of cigarette smoking identified at the time of the pre-operative patient assessment should be taken into account as a potential negative predictive factor for potency recovery after surgery [12].

Surgical technique

In suitable candidates, radical excision of the prostate should be performed with the objective of achieving total cancer control, i.e., removing all cancer present in the prostatic tissue, while maintaining the integrity of the anatomical structures on which urinary continence and erectile function are based. Namely, the corpora cavernosa receive the innervation responsible for erections through the cavernosal nerves which branch out from the pelvic plexus. This latter structure is located adjacent to the tip of the seminal vesicles on the anterolateral wall of the rectum and may be damaged during radical prostatectomy. The cavernous nerves course adjacent to small vessels forming the so-called neurovascular bundle along the posterolateral margin of the prostate, bilaterally, and are located between the visceral layer of the endopelvic fascia and the prostatic fascia. The neurovascular bundle is located at the 5- and 7 o'clock positions at the level of the membranous urethra and, after piercing the urogenital diaphragm, it enters the corpora cavernosa where it innervates the smooth muscle cells of the penile vessel walls and sinusoids [15]. Based on the better understanding of the surgical anatomy of the prostate, Walsh has clearly described the technique of anatomic radical prostatectomy in a step-by-step fashion [15]. The use of a frontal xenon light and of magnifying loupes significantly improves vision during the procedure and the authors of this review feel that this armamentarium should be always used when performing nerve sparing radical prostatectomies. Some crucial surgical steps deserve particular attention. Following pelvic lymphadenectomy, the endopelvic fascia is incised to allow the dissection of the prostatic apex. Care should be taken to ligate small branches of pudendal vessels which are located just underneath the endopelvic fascia in the area of the prostatic apex; cautery should not be used to secure these vessels in order to avoid damage to pudendal nerve branches also located there. Ligation of the Santorini's plexus is a fundamental step in the procedure as it must guarantee perfect hemostasis in order to obtain the best visualization of the surgical field during the excision of the prostate. Walsh has suggested to ligate the Santorini's plexus distally first and to subsequently divide it with scissors; control of the proximal stump of the venous plexus is then achieved with a V-shaped suture which avoids a central retraction of the neurovascular bundles thus facilitating the nerve sparing procedure. After controlling and dividing the Santorini's plexus, the visceral layer of the endopelvic fascia should be bluntly incised on the lateral side of the prostate to allow for the gentle lateral displacement of the neurovascular bundles which is obtained with the use of small sponge sticks. If this manoeuvre is performed correctly, the neurovascular bundles become clearly visible along their entire course from the membranous urethra and the seminal vesicles in most patients. It is of mandatory importance to avoid the use of the cautery during every step of the prostatic excision as this can facilitate a definitive damage of the neurovascular bundles. As the pelvic plexus is located adjacent to the posterolateral tips of the seminal vesicles, particular care should be taken while excising them. The possible advantage of partially excising the seminal vesicles to reduce the risk of damaging the pelvic plexus has been reported [16]. Hemostasis must be accurate and this is usually obtained with small sized clips. When pelvic hematomas occur it is advisable to drain them surgically as the fibrosis occurring after the slow spontaneous resolution of any blood collection significantly lengthens the duration of cavernosal neuropraxia. Reviewing the videotapes of one's own nerve sparing radical prostatectomies and comparing the surgical technique with the results seen in terms of margins status, urinary continence and erectile function may be of significant importance as it allows to identify the parts of the operation which are at higher risks for surgical errors from the surgeon [17].

Pathophysiology of erectile dysfunction following nerve sparing radical prostatectomy

Patients undergoing nerve sparing radical prostatectomy often experience impairment of erections in the early postoperative period. This has been related to the development of neuropraxia which is believed to be caused by some damage to the cavernosal nerves, which inevitably occurs during the excision of the prostate even in the hands of most experienced surgeons. Absence of early postoperative erections is associated with poor corporeal oxygenation which may facilitate the development of corporeal fibrosis which ultimately leads to veno-occlusive dysfunction [18]. Recently the role of apoptosis has been considered in the pathophysiology of post prostatectomy ED. Apoptosis, or programmed cell death, is essential for the normal development of multicellular organisms as well as for physiological cell turnover. Morphologically this phenomenon is characterized by chromatin condensation, membrane blabbing and cell volume loss. In the penis chronic hypoxia and denervation have been shown to stimulate apoptosis: it is possible that cellular apoptosis leads to increased deposition of connective tissue that may finally lead to a decrease in penile distensibility [19–22]. Recently, User et al. [23] have elegantly elucidated the role of apoptosis in the patho-physiology of post-prostatectomy erectile dysfunction in the rat model. Postpubertal rats were randomized to bilateral or unilateral cavernous nerve transection *versus* a sham operation. At different time intervals following the procedure, penile wet weight, DNA content and protein content were measured. Tissue sections of the penis were stained for apoptosis and the apoptotic index was calculated. Finally, staining for endothelial and smooth muscle cells was done to identify the apoptotic cell line. Wet weight of the denervated penis was significantly decreased after bilateral cavernous neurotomy while unilateral cavernous neurotomy allowed much greater preservation of penile weight. DNA content was also significantly reduced in bilaterally denervated penes while no difference was found between unilaterally denervated penis and controls. Bilateral cavernous neurotomy induced significant apoptosis which peaked on postoperative day two. In addition, it was found that most apoptotic cells were located just beneath the tunica albuginea of the corpus cavernosum, i.e., the anatomical area where the subtunical venular plexus is located. Finally the authors found that apoptotic cells were smooth muscle cells and not endothelial cells. The subsequent hypothesis suggested by the authors was that the bilateral injury to the cavernous nerves may induce significant apoptosis of smooth muscle cells, particularly in the subtunical area, thus causing an abnormality of the veno-occlusive mechanism of the corpus cavernosum. Apoptosis seems to play a role also in the genesis of ED seen in the aging population (which is clearly of crucial importance in the radical prostatectomy patient group) as it has been shown that anti-apoptotic genes and proteins are expressed in young rats but not in aging rats [19]. These findings on apoptosis following radical prostatectomy confirm the known fact that most patients reporting postoperative ED do develop massive corporeal venous leaks on the long term [24]. However, a reduction of the arterial inflow to the corpora cavernosa in patients with post-prostatectomy erectile dysfunction has been reported by several authors as a significant etiological co-factor [25, 26]. In conclusion, the postoperative combination of reduced penile arterial inflow and excessive venous outflow due to the apoptosis-induced damage of the veno-occlusive mechanism, lead to reduced oxygen transport and increased production of transforming growth factor beta. This is subsequently causing significant tissue damage, i.e., increased corporeal fibrosis, which not only is at the root of the known penile hemodynamic abnormality but is also probably causing the postoperative decrease of penile length recently reported in an interesting study [27].

Pharmacological prophylaxis and treatment of postoperative erectile dysfunction

Prophylaxis

The better understanding of pathophysiology of post-prostatectomy ED including the concept of tissue damage induced by poor corporeal oxygenation paved the way to the application of pharmacological regimens aimed at improving early postoperative corporeal blood filling. Montorsi et al. [28] showed that by using intracorporeal injections of alprostadil early after a bilateral nerve sparing radical prostatectomy the rate of recovery of spontaneous erections was significantly higher than observation alone. In the author's experience, alprostadil injection therapy should be started as soon as possible after the procedure, usually at the end of the first postoperative month. The initial dose is alprostadil 5 μg. Patients should be instructed to use injections 2–3 times per week in order to obtain penile tumescence; injections are not necessarily associated with sexual intercourse but is clear that if the patient desires, he may be able to resume satisfactory sexual intercourse with penetration as soon as he is able to identify the correct dosing of the drug. Brock et al. [29] have also shown that the continuous use of intracavernous alprostadil injection therapy was able to significantly improve penile haemodynamics and return of spontaneous erections (either partial or total) in patients with arteriogenic ED, thus confirming a potential curative role of this therapeutic modality in selected patients. The early postoperative use of intracavernosal injection therapy may also exert a significant psychological role. Commonly after a nerve sparing radical prostatectomy with the slow return of spontaneous erections, a dysfunctional sexual dynamic may develop in couples. The patient withdraws sexually as he is increasingly discouraged with his lack of erectile function, which is a constant reminder of his cancer. The female partner, relieved that the patient has survived the surgery, may be satisfied with his companionship and is not anxious to upset him by making sexual overtures that may frustrate him. Successful intracorporeal injection therapy early after radical prostatectomy may contribute to break this negative cycle [12].

The advent of PDE5-I in the treatment of ED has clearly revolutionized the management of this medical condition. This class of agents acts within the smooth muscle cell by inhibiting the enzyme phosphodiesterase type 5 which naturally degrades cyclic guanosin-mono-phosphate (cGMP), an intracellular nucleotide which acts as second messenger in the process of smooth muscle cell relaxation. The cascade of intracellular events which leads to the relaxation of the smooth muscle cell is initiated by the release of nitric oxide (NO) which follows sexual stimulation. Both intracorporeal cavernous nerve terminals and endothelium release NO, which as a gas diffuses into the smooth muscle cells and activate the enzyme guanylate cyclase, which ultimately catalyzes the reaction from guanosintriphosphate to cGMP. Increased levels of cGMP lead to the activation of cGMP-specific protein kinases which activate further intracellular events leading to the final reduction of intracellular calcium, this being associated with smooth muscle cell relaxation. At present, sildenafil, tadalafil and vardenafil are approved for clinical use in the European Union (EU) and have also been utilized as a post-prostatectomy ED treatment. The rationale for the use of these drugs as prophylaxis is not yet well understood. The basic concept would be to administer a PDE5-I at bedtime in order to facilitate the occurrence of nocturnal erections, which are believed to have a natural protective role on the baseline function of the corpora cavernosa. Montorsi et al. showed that when sildenafil 100 mg is administered at bedtime in patients with ED of various etiologies, the overall quality of nocturnal erections as recorded with the RigiScan device is significantly better than those obtained after the administration of placebo [30]. Padma Nathan et al. [31] recently reported on the prospective administration of sildenafil 50 and 100 mg *versus* placebo, daily and at bedtime, in patients undergoing bilateral nerve sparing radical prostatectomy who were potent preoperatively. Four weeks after surgery patients were randomized to sildenafil or placebo, which were carried on for 36 weeks. Eight weeks after discontinuation of treatment erectile function was assessed with the question "Over the past four weeks, have your erections been good enough for satisfactory sexual activity?" and by IIEF and nocturnal penile tumescence assessments. Responders were defined as those having a combined score of ≥8 for IIEF questions three and four and a positive response to the above question. 27% of the patients receiving sildenafil were responders, i.e., demonstrated return of spontaneous normal erectile function compared to 4% in the placebo group (p = 0.0156). Postoperative nocturnal penile tumescence (NPT) assessments were supportive. We believe that in the hands of experienced surgeons a 27% overall rate of return to normal erectile function after a bilateral nerve sparing procedure is far from being impressive but the important message from this study is that daily bedtime administration of sildenafil 50 or 100 mg after this procedure should be able to improve every surgeon's baseline results. Although similar data are not available yet for tadalafil and vardenafil we believe there is no reason not to expect similar findings with these drugs. In practical terms, we feel that the need for postoperative prophylactic therapy should be discussed with the patient when counseling him about radical prostatectomy as one of the treatment options for prostate cancer. Patients must have the chance to guide their choice also by being informed on the pharmacological approaches needed to recover a normal postoperative erectile function.

Treatment

Phosphodiesterase type 5 inhibitors

Phosphodiesterase type 5 inhibitors have acquired an established role in the treatment of post prostatectomy erectile dysfunction. As the mechanism of action of this class of drugs implies the presence of NO within the corporeal smooth muscle cells, only patients undergoing a nerve sparing procedure should be expected to respond to these agents. Sildenafil is the drug that has been studied more extensively in this patient subgroup as it has been on the market worldwide since 1998. In general terms, it is now known that sildenafil allows to obtain the best results in young patients, the less than 60-year-old age group being the best responding, in patients treated with a bilateral nerve sparing procedure and in patients who show some degree of spontaneous postoperative erectile function. Sildenafil is usually administered at the largest available dose although it is common experience to have post-prostatectomy patients responding also to the 25 and 50 mg doses. Typically the response to sildenafil has been shown to improve as time passes after the procedure: best results are seen from 12–24 months postoperatively [5, 32–36]. Unfortunately, no data coming from multicenter, randomized, placebo controlled trials assessing sildenafil in patients undergoing nerve sparing radical prostatectomy are available today. It has been suggested that the early postoperative prophylactic administration of alprostadil injections allowed to obtain a subsequent better response rate to sildenafil with lower doses of the drug being necessary [37]. Vardenafil has been tested in patients treated with ED following a uni- or bilateral nerve sparing prostatectomy in a multicenter, prospective, placebo controlled, randomized study done in the US and Canada. This was a 12-week parallel arm study comparing placebo to vardenafil 10 and 20 mg. In this study sildenafil failures were excluded. 71% and 60% of patients treated with a bilateral nerve sparing procedure reported an improvement of erectile function following the administration of vardenafil 20 mg and 10 mg, respectively. A positive answer to SEP 2 question ("were you able to insert your penis into your partner's vagina?") was seen in 47% and 48% of patients using vardenafil 10 and 20 mg, respectively. A positive answer to the more challenging SEP 3 question ("did your erections last enough to have successful intercourse?") was seen in 37% and 34% of patients using vardenafil 10 and 20 mg, respectively [38] Tadalafil was also evaluated in a large multicenter trial conducted in Europe and in the US involving patients with erectile dysfunction following a bilateral nerve sparing procedure. 71% of patients treated with tadalafil 20 mg reported the improvement of their erectile function as compared to 24% of those treated with placebo ($p < 0.001$). The erectile function domain score of the IIEF was significantly higher after treatment with tadalafil 20 mg compared to placebo (21.0 *versus* 15.2; $p < 0.001$) and this difference was clinically significant [13]. Tadalafil 20 mg allowed to achieve a 52% rate of successful intercourse attempts which was significantly higher than the 26% obtained with a placebo ($p < 0.001$) [39]. The adverse event profile of the three PDE5-I has been very similar also in this patient population and the authors of this review feel that discontinuation from treatment with one of the PDE5-I is usually caused by lack of efficacy while tolerability is overall more than satisfactory. We believe that there is a need to obtain comparative data between the three drugs in terms of efficacy, safety, tolerability and overall patient preference in this challenging patient population. Well designed head-to-head trials are ongoing at present and this information will become available soon.

Other medical treatments
Patients, either non-responding to or who cannot use PDE5-I are typical candidates for second-line pharmacological treatment, which currently includes the intraurethral and the intracorporeal administration of alprostadil. The most recent information on the use of intraurethral alprostadil use suggests that its combination with oral sildenafil may salvage a significant proportion of sildenafil failures [40, 41]. Intracorporeal injection therapy with alprostadil is effective in the majority of post-prostatectomy patients, regardless of the status of their cavernosal nerves. Recently, Gontero et al. have shown that the best clinical and hemodynamic response with alprostadil injection therapy is seen a month after the procedure but that the attrition rate for this approach is high when it is started early after surgery. They proposed that in order to optimize long-term success with intracavernous injections of alprostadil, this treatment should be started three months after surgery [42]. In the experience of the authors of this review, the use of alprostadil monotherapy should be matched against the use of the combination of alprostadil, papaverine and phento-

lamine. Patients treated with the three drug combination obtain very high response rates while reporting penile pain, the typical adverse event related to alprostadil, only rarely [43]. The major disadvantage of the use of the three drug combination is caused by its non-availability on the market which obliges the urologist to prepare the solution on his own and then distribute it to the patient. This off label use requires the approval of the Ethics Committee of the hospital and a signed informed consent from the patient.

Cavernous nerve reconstruction

Cavernous nerve reconstruction has generated great interest recently to preserve erectile function after radical prostatectomy. The interest is guided by the premise that nerve grafts provide a scaffold for autonomic nerve regrowth and reconnection with targets that mediate erectile function. Original preclinical studies by Quinlan et al. [44] and Ball et al. [45] using the genitofemoral nerve as a replacement for resected cavernous nerves in rats introduced the possibility of interposition nerve grafting for the recovery of erectile function after injury of the penile innervation. The surgical application was extended to the human level in the context of radical prostatectomy, initially by Walsh who used the genitofemoral nerve as an autologous nerve graft [46], and subsequently by Kim et al. [47] who applied the sural nerve as a conduit.

Kim et al. described definite promise for cavernous nerve interposition grafting based on their success [47–52]. Among several reports by this group, they described their extended experience involving 28 men undergoing bilateral sural nerve grafting at radical prostatectomy (with a mean follow up of 23 months), of whom 6 (26%) of 23 men completing IIEF questionnaires had spontaneous, medically unassisted erections sufficient for sexual intercourse and an IIEF erectile function domain score of 20 [51]. Their inclusion of four additional men whose partial erections were enhanced sufficiently to allow sexual intercourse using sildenafil resulted in an overall potency rate of 43% (10 of 23) [51]. They also identified the benefit of unilateral sural nerve grafting based on a 78% erectile function rate with this treatment (compared with the 30% rate without grafting), which approximated the level of recovery observed in their hands for men having undergone bilateral nerve sparing surgery (79%) [50, 51].

Other groups have also described their early experiences with sural nerve grafting after radical prostatectomy. Anastasiadis et al. described 4 (33%) of 12 men who achieved spontaneous erections sufficient for intercourse based on an IIEF erectile function domain score of 20 (with a mean follow up of 16 months) after unilateral nerve grafting (including one man recovering erections who required bilateral cavernous nerve excision) [53]. Chang et al. determined that 18 (60%) of 30 patients who underwent bilateral autologous sural nerve grafting achieved spontaneous erectile ability and 13 (43%) of these 30 were able to have intercourse (with a mean follow up of 23 months) [54].

The evolution of the procedure has related to the application of a variety of technical modifications. One such modification is the use of electrical stimulation to confirm the function and location of the recipient nerve [50–54]. Additional recommendations include the use of microsurgical instruments and loupe magnification and attention to such procedural details as maintenance of hemostasis within the surgical field and completion of grafting without tension. The feasibility of sural nerve grafting by a laparoscopic approach has also been demonstrated [55].

Before assigning the fundamental utility of cavernous nerve interposition grafting as for any innovation, the procedure must withstand a critical process of evaluation. Clearly, advantages of sural nerve grafting include the autogenous basis for the nerve conduit, which confers biocompatibility without apparent immunogenicity, as well as support from earlier demonstrations that autonomic nerve replacement grafting succeeds [56–58]. The reportedly negligible functional compromise or consequence from the donor site also represents a favorable consideration. However, concerns persist along the primary grounds that the procedure may be technically impracticable and otherwise infrequently useful. The first consideration relates to the difficulty in identifying and performing grafting in humans in which the cavernous nerve is contained within a heterogeneous structure of nerve fibers and blood vessels unlike the rat in which the cavernous nerve is easily identifiable as a large, discrete cable-like structure. Of greater concern is the issue that the majority of candidates for radical prostatectomy in the modern era of early prostate cancer detection are likely to be satisfactorily treated with nerve-sparing techniques that do not compromise the intent of cancer control. On this basis, the widespread applicability of this procedure may be questioned. Quite likely, the candidate selected for nerve grafting will have high stage disease that is associated with a low probability of sexual function recovery after surgery [46] and likely will require adjuvant therapy, further reducing the likelihood of the graft preserving erectile function [53]. A fair statement at this time may well be that this innovation requires further investigation particularly through clinical trials, before establishing its optimal role.

Neuroprotection

Neuroprotective/neurotrophic interventions have also rapidly gained attention as an approach to promote or restore cavernous nerve function in the face of nerve injury, particularly after radical prostatectomy. Early considerations for therapeutic neurogenesis, which implies axonal growth and regeneration, involved rat studies in which neurotrophic growth factors, e.g., nerve growth factor (NGF), acidic fibroblast growth factor (FGF), were applied in combination with nerve interposition grafts to enhance erectile function recovery [59, 60]. Early evaluations of growth factors in the rat penis led to support for basic FGF as a predominant neurotrophic factor [61], while a host of other growth factors have also been identified in this organ [62].

At this time, multiple factors have been described as potentially active neurotrophic effectors in the penis. These include growth hormone [63], neurturin [64], immunophilins [65], Sonic hedgehog protein [66], vascular endothelial growth factor [67], and prostaglandin E1 [68]. Beyond their molecular characterizations and evidence of their neurite outgrowth properties in the penis, several of these factors have been further shown to have relevance according to physiologic studies. For instance, non-immunosuppressant immunophilin ligands [69], brain-derived neurotrophic factor (BDNF) [70], and insulin-like growth factor-1 [71] have been shown to recover cavernous nerve function for improved erection responses after various forms of nerve injury in rat models.

Various other neurobiologic concepts have also been preliminarily investigated as potential approaches to recover cavernous nerve function. Delivery methods for neurotrophic factors have combined gene therapy, tissue engineering, and tissue reconstitution techniques. Adeno-associated virus gene therapy has been successfully applied to deliver BDNF to the penis and recover erections in cavernous nerve-injured rats [72, 73]. Similar success for delivery of neurotrophic factors has been shown using a herpes simplex virus vector [74]. Plasmid DNA preparations have been used to deliver the antisense of protein inhibitor of NOS (nitric oxide synthase), yielding improved erection presumably by promoting the action of the major neuromediator of penile erection NO [75]. Intracorporeal autotransplantation of pelvic ganglia in rats has shown promise as a therapy for neurogenic ED [76]. Muscle-derived stem cells injected at the site of cavernous nerve injury in rats was also recently described to recover erectile function [77]. Novel conduits under investigation in rats to serve as cavernous nerve replacements include Schwann cells transplanted into a silicon tube nerve guide [78] and acellular nerve matrices [79].

As the field of therapeutic neurogenesis evolves, interest is high to bring its applications for penile innervation to a clinical level following radical prostatectomy. While many burgeoning avenues have generated a great deal of excitement, they have been explored really at only the earliest development stages. Among prospects having imminent clinical roles, immunophilin ligands offer a most promising direction. Given this possibility, further description of their discovery and preclinical investigation seems appropriate at this time.

The field of immunophilin research applicable to neural roles evolved from the discovery that receptor proteins for such immunosuppressant drugs as FK506 and cyclosporin A, called immunophilins, are expressed at a 50-fold greater concentration in the nervous system than in immune cells [80] and exert diverse neural functions including regulation of NO neurotoxicity, neurotransmitter release, and mediation of neurotrophic influences [81, 82]. The impetus to evaluate the neuroprotective effects of immunophilin ligands in the penile innervation after nerve injury followed reports that the prototype immunophilin ligand FK506 promotes both nerve regeneration and functional recovery *in vivo* using animal models of crush injury of sciatic nerves [83, 84] and neuroprotection after injury to other nerve structures such as the optic and facial nerves [85–87].

In an initial report using a rat model of cavernous nerve injury developed to resemble the partial nerve injury associated with nerve-sparing radical prostatectomy, FK506 was shown to be an effective neuroprotective agent for preserving erectile function and preventing cavernous nerve degeneration [65]. Subsequent studies confirmed that FK506 binding protein 12 (FKBP-12), a prominent member of the immunophilin family of receptor proteins, is expressed in the rat penile innervation [88]. Additionally, both FK506 and GPI1046, a FK506 derivative that is non-immunosuppressive exhibited neuroregenerative effects on penile innervation and erectile tissue histology in rats with extensive cavernous nerve injury [69]. Preliminary study has further confirmed that non-immunosuppressant FK506 derivatives do not promote growth of prostate cancer cells *in vitro*, addressing concerns of the therapy having mitogenic effects on prostate tissue (Burnett AL, *personal communication*).

Based on the current progress of the field involving immunophilin ligands to preserve the penile innervation, clinical trials in humans following radical prostatectomy are now underway to investigate whether these agents will be pharmacotherapeutically effective clinically. The trial, sponsored by Guilford Inc, involves the administration of a non-immunosuppressant ligand derivative, hence avoiding adverse treatment effects of immunosuppressant compounds, formulated as an oral preparation at this time to pre-operatively potent men undergoing bilateral nerve-sparing radical prostatectomy. By clinical trial design, specifications of the investigation include multicenter patient enrolment, randomized, placebo-control methodology, and use of currently validated self-report instruments to assess erectile function recovery. It is anticipated that the results of the trial will show a more rapid and complete recovery of penile erection with use of the therapy than in its absence, while it may also facilitate responses to other currently available ED treatments such as PDE5 inhibitor therapy. The study represents a pioneering investigation in therapeutic neurogenesis and will certainly model other similar studies to come. Further basic investigation in the field of immunophilin research will pertain to characterizing immunophilin receptor proteins and their signaling mechanisms responsible for neuroprotection and neuroregeneration in the penis more fully and evaluating the possible interactions between immunophilins and neurotrophic systems operating in the penis such as the NO signal transduction pathway and conventional neurotrophins, e.g., nerve growth factor.

The intensity of research activity in this field suggests that therapeutic neurogenesis is rapidly moving forward as a highly promising area for treating the ED that commonly follows radical prostatectomy and conceivably other presentations of neurogenic erectile dysfunction as well. While the extent of study in this area has been essentially preclinical thus far, there is major enthusiasm to bring novel applications to humans clinically in the near future. For radical prostatectomy, such interventions are useful whether nerve sparing or non-nerve sparing techniques are applied. Hence, their relevance in the modern management of this disease process is high, with applicability not just for the surgical management of the disease process but also for other forms of treatment such as pelvic irradiation in which cavernous nerve injury is known to occur. Further advances will come with ongoing investigation of molecular mechanisms associated with penile neuropathy after cavernous nerve injury and cavernous nerve neurogenesis, in combination with timely applications of neurobiologic strategies relevant to this area.

Conclusions

Radical prostatectomy is an increasingly performed procedure in patients with prostate cancer. As the mean age of this patient subgroup is progressively declining due to the advent of prostate cancer screening programs, the demand for optimal postoperative quality of life is becoming more important. Erectile function can be preserved in patients undergoing a nerve sparing radical prostatectomy provided that the patient is rigorously selected prior to surgery and that an anatomic radical prostatectomy following the most modern templates is precisely performed. Pharmacological prophylaxis, either with oral or intracavernosal drugs, may potentially have a significantly expanding role in the future strategies aimed at preserving postoperative erectile function. The advent of the on demand use of PDE5-I has clearly improved the overall postoperative potency rate also in this challenging subgroups of patients. Future management avenues including cavernous nerve reconstruction and neuroprotection strategies are based on interesting rationales but need to undergo the test of time.

This manuscript was partially read as a State of the Art Lecture during the 2003 annual Meeting of the European Association of Urology held in Madrid, Spain.

References

1 Holmberg L, Bill-Axelson A, Helgesen F, Salo JO, Folmerz P, Haggman M, Andersson SO, Spangberg A, Busch C, Nordling S et al; Scandinavian Prostatic Cancer Group Study Number 4 (2002) A randomised trial comparing radical prostatectomy with watchful waiting in early prostate cancer. *N Engl J Med* 347: 781–789
2 Walsh PC, Donker PJ (1982) Impotence following radical prostatectomy: insight into etiology and prevention. *J Urol* 128: 492–497
3 Walsh PC, Marschke P, Ricker D, Burnett AL (2000) Patient-reported urinary continence and sexual function after anatomic radical prostatectomy. *Urology* 55: 58–61
4 Walsh PC (2000) Radical prostatectomy for localized prostate cancer provides durable cancer control with excellent quality of life: a structured debate. *J Urol* 163: 1802–1807
5 Rabbani F, Stapleton AM, Kattan MW, Wheeler TM, Scardino PT (2000) Factors predicting recovery of erections after radical prostatectomy. *J Urol* 164: 1929–1934
6 Stanford JL, Feng Z, Hamilton AS, Gilliland FD, Stephenson RA, Eley JW, Albertsen PC, Harlan LC, Potosky AL (2000) Urinary and sexual function after radical prostatectomy for clinically localized prostate cancer: The Prostate Cancer Outcomes Study. *JAMA* 283: 354–360
7 Mulcahy JJ (2000) Erectile function after radical prostatectomy. *Semin Urol Oncol* 18: 71–75
8 Lepor H (2003) Practical considerations in radical retropubic prostatectomy. *Urol Clin North Am* 30: 363–368

9. Montorsi F, Salonia A, Zanoni M, Colombo R, Pompa P, Rigatti P (2001) Counselling the patient with prostate cancer about treatment-related erectile dysfunction. Curr Opin Urol 11: 611–617
10. Meuleman EJ, Mulders PF (2003) Erectile function after radical prostatectomy: a review. Eur Urol 43: 95–101
11. Aus G, Abbou CC, Pacik D, Schmid HP, Van Poppel H, Wolff JM, Zattoni F (2003) Guidelines on prostate cancer. EAU Guidelines 2004
12. McCullough AR (2001) Prevention and management of erectile dysfunction following radical prostatectomy. Urol Clin North Am 28: 613–627
13. Rosen RC, Cappelleri JC, Gendrano N 3rd (2002) The International Index of Erectile Function (IIEF): a state-of-the-science review. Int J Impot Res 14: 226–244
14. Wespes E (2000) Erectile dysfunction in the ageing man. Curr Opin Urol 10: 625–628
15. Walsh PC (2002) Anatomic radical retropubic prostatectomy. In: Walsh PC, Retik AB, Vaughan ED Jr., Wein AJ (eds): *Campbell's Urology*, 8th ed. WB Saunders, Philadelphia, 3107–3129
16. Sanda MG, Dunn R, Wei JT, Hoag L, Montie JE, Arbor A (2003) Sexual function recovery after prostatectomy based on quantified pre-prostatectomy sexual function and use of nerve-sparing and seminal-vesicle-sparing surgical technique. J Urol 169 (Suppl): 181
17. Walsh PC, Marschke P, Ricker D, Burnett AL (2000) Use of intraoperative video documentation to improve sexual function after radical retropubic prostatectomy. Urology 55: 62–67
18. Moreland RB (1998) Is there a role of hypoxemia in penile fibrosis: a viewpoint presented to the Society for the Study of Impotence. Int J Impotence Res 10: 113–120
19. Yamanaka M, Shirai M, Shiina H, Shirai M, Tanaka Y, Fujime M, Okuyama A, Dahiya R (2002) Loss of anti-apoptotic genes in aging rat crura. J Urol 168: 2296–2300
20. Yao KS, Clayton M, O'Dwyer PJ (1995) Apoptosis in human adenocarcinoma HT29 cells induced by exposure to hypoxia. J Natl Cancer Inst 158: 656–659
21. Chung WS, Park YY, Kwon SW (1997) The impact of aging on penile hemodynamics in normal responders to pharmacological injection: a Doppler sonographic study. J Urol 157: 2129–2131
22. Klein LT, Miller MI, Buttyan R, Raffo AJ, Burchard M, Devris G, Cao YC, Olsson C, Shabsigh R (1997) Apoptosis in the rat penis after penile denervation. J Urol 158: 626–630
23. User HM, Hairston JH, Zelner DJ, McKenna KE, McVary KT (2003) Penile weight and cell subtype specific changes in a post-radical prostatectomy model of erectile dysfunction. J Urol 169: 1175–1179
24. De Luca V, Pescatori ES, Taher B, Zambolin T, Giambroni L, Frego E, Cosciani Cunico S (1996) Damage to the erectile function following radical pelvic surgery: prevalence of veno-occlusive dysfunction. Eur Urol 29: 36–40
25. Mulhall JP, Slovick R, Hotaling J, Aviv N, Valenzuela R, Waters WB, Flanigan RC (2002) Erectile dysfunction after radical prostatectomy: hemodynamic profiles and their correlation with the recovery of erectile function. J Urol 167: 1371–1375
26. Mulhall JP, Graydon RJ (1996) The hemodynamics of erectile dysfunction following nerve-sparing radical prostatectomy. Int J Impot Res 8: 91–94
27. Savoie M, Kim SS, Soloway MS (2003) A prospective study measuring penile length in men treated with radical prostatectomy for prostate cancer. J Urol 169: 1462–1464
28. Montorsi F, Guazzoni G, Strambi LF, Da Pozzo LF, Nava L, Barbieri L, Rigatti P, Pizzini G, Miani A (1997) Recovery of spontaneous erectile function after nerve-sparing radical retropubic prostatectomy with and without early intracavernous injections of alprostadil: results of a prospective, randomised trial. J Urol 158: 1408–1410
29. Brock G, Tu LM, Linet OI (2001) Return of spontaneous erection during long-term intracavernosal alprostadil (Caverject) treatment. Urology 57: 536–541
30. Montorsi F, Maga T, Strambi LF, Salonia A, Barbieri L, Scattoni V, Guazzoni G, Losa A, Rigatti P, Pizzini G (2000) Sildenafil taken at bedtime significantly increases nocturnal erections: results of a placebo-controlled study. Urology 56: 906–911
31. Padma-Nathan E, McCullough AR, Giuliano F, Toler SM, Wohlhuter C, Shpilsky AB (2003) Postoperative nightly administration of sildenafil citrate significantly improves the return of normal spontaneous erectile function after bilateral nerve-sparing radical prostatectomy. J Urol 4 (Suppl): 375
32. Zippe CD, Jhaveri FM, Klein EA, Kedia S, Pasqualotto FF, Kedia A, Agarwal A, Montague DK, Lakin MM (2000) Role of Viagra after radical prostatectomy. Urology 55: 241–245
33. Lowentritt BH, Scardino PT, Miles BJ, Orejuela FJ, Schatte EC, Slawin KM, Elliott SP, Kim ED (1999) Sildenafil citrate after radical retropubic prostatectomy. J Urol 162: 1614–1617

34 Hong EK, Lepor H, McCullough AR (1999) Time dependent patient satisfaction with sildenafil for erectile dysfunction (ED) after nerve-sparing radical retropubic prostatectomy (RRP). *Int J Impot Res* 11 (Suppl 1): S15–22
35 Feng MI, Huang S, Kaptein J, Kaswick J, Aboseif S (2000) Effect of sildenafil citrate on post-radical prostatectomy erectile dysfunction. *J Urol* 164: 1935–1938
36 Blander DS, Sanchez-Ortiz RF, Wein AJ, Broderick GA (2000) Efficacy of sildenafil in erectile dysfunction after radical prostatectomy. *Int J Impot Res* 12: 165–168
37 Montorsi F, Salonia A, Barbieri L, Maga T, Zanoni M, Raber M, Bua L, Cestari A, Guazzoni G, Rigatti P (2002) The subsequent use of I.C. alprostadil and oral sildenafil is more efficacious than sildenafil alone in nerve sparing radical prostatectomy. *J Urol* 167: 279 (abstract 1098)
38 Brock G, Terry T, Monica S, Vardenafil PROSPECT Group Canada (2002) Efficacy and tolerability of vardenafil in men with erectile dysfunction following radical prostatectomy. *Eur Urol* 1 (Suppl): 152
39 Montorsi F, McCullough A, Brock GB, Broderick G, Ahuja S, Hoover A, Novack D, Murphy A, Varanese L, Tadalafil in the Treatment of Erectile Dysfunction following Bilateral Nerve-Sparing Radical Retropubic Prostatectomy: A Randomized, Double-blind, Placebo-controlled trial. *J Urol*; *in press*
40 Nehra A, Blute ML, Barrett DM, Moreland RB (2002) Rationale for combination therapy of intraurethral prostaglandin E(1) and sildenafil in the salvage of erectile dysfunction patients desiring noninvasive therapy. *Int J Impot Res* 14 (Suppl 1): S38–42
41 Raina R, Oder M, Afarwal A, Zippe CD (2003) Combination therapy: muse enhances sexual satisfaction in sildenafil citrate failures following radical prostatectomy (RP). *J Urol* 169;4 (Suppl): 354
42 Gontero P, Fontana F, Bagnasacco A, Panella M, Kocjancic E, Pretti G, Frea B (2003) Is there an optimal time for intracavernous prostaglandin E1 rehabilitation following nonnerve sparing radical prostatectomy? Results from a hemodynamic prospective study. *J Urol* 169: 2166–2169
43 Montorsi F, Guazzoni G, Bergamaschi F, Dodesini A, Rigatti P, Pizzini G, Miani A (1993) Effectiveness and safety of multidrug intracavernous therapy for vasculogenic impotence. *Urology* 42: 554–558
44 Quinlan DM, Nelson RJ, Walsh PC (1991) Cavernous nerve grafts restore erectile function in denervated rats. *J Urol* 145: 380–383
45 Ball RA, Richie JP, Vickers MA, Jr. (1992) Microsurgical nerve graft repair of the ablated cavernosal nerves in the rat. *J Surg Res* 53: 280–286
46 Walsh PC (2001) Nerve grafts are rarely necessary and are unlikely to improve sexual function in men undergoing anatomic radical prostatectomy. *Urology* 57: 1020–1024
47 Kim ED, Scardino PT, Hampel O, Mills NL, Wheeler TM, Nath RK (1999) Interposition of sural nerve restores function of cavernous nerves resected during radical prostatectomy. *J Urol* 161: 188–192
48 Kim ED, Scardino PT, Kadmon D, Slawin K, Nath RK (2001) Interposition sural nerve grafting during radical retropubic prostatectomy. *Urology* 57: 211–216
49 Kim ED, Nath R, Kadmon D, Lipshultz LI, Miles BJ, Slawin KM, Tang HY, Wheeler T, Scardino PT (2001) Bilateral nerve graft during radical retropubic prostatectomy: 1-year followup. *J Urol* 165: 1950–1956
50 Scardino PT, Kim ED (2001) Rationale for and results of nerve grafting during radical prostatectomy. *Urology* 57: 1016–1019
51 Kim ED, Nath R, Slawin KM, Kadmon D, Miles BJ, Scardino PT (2001) Bilateral nerve grafting during radical retropubic prostatectomy: extended follow-up. *Urology* 58: 983–987
52 Canto EI, Nath RK, Slawin KM (2001) Cavermap-assisted sural nerve interposition graft during radical prostatectomy. *Urol Clin North Am* 28: 839–848
53 Anastasiadis AG, Benson MC, Rosenwasser MP, Salomon L, El-Rashidy H, Ghafar MA, McKiernan JM, Burchardt M, Shabsigh R (2003) Cavernous nerve graft reconstruction during radical prostatectomy or radical cystectomy: safe and technically feasible. *Prostate Cancer Prostatic Dis* 6: 56–60
54 Chang DW, Wood CG, Kroll SS, Youssef AA, Babaian RJ (2003) Cavernous nerve reconstruction to preserve erectile function following non-nerve-sparing radical retropubic prostatectomy: a prospective study. *Plast Reconstr Surg* 111: 1174–1181
55 Turk IA, Deger S, Morgan WR, Davis JW, Schellhammer PF, Loening SA (2002) Sural nerve graft during laparoscopic radical prostatectomy. Initial experience. *Urol Oncol* 7: 191–194
56 Hammerschlag PE (1999) Facial reanimation with jump interpositional graft hypoglossal facial

anastomosis and hypoglossal facial anastomosis: evolution in management of facial paralysis. *Larynogoscope* 109: 1–23
57. Hogikyan ND, Johns MM, Kileny PR, Urbanchek M, Carroll WR, Kuzon WM, Jr. (2001) Motion-specific laryngeal reinnervation using muscle-nerve-muscle neurotization. *Ann Otol Rhinol Laryngol* 110: 801–810
58. Golub DM, Kheinman FB, Novikov II, Archakova LI, Danilenko RV (1979) Reinnervation of internal organs and vessels by the neuropexy technic. *Arkh Anat Gistol Embriol* 76: 5–16
59. Burgers JK, Nelson RJ, Quinlan DM, Walsh PC (1991) Nerve growth factor, nerve grafts and amniotic membrane grafts restore erectile function in rats. *J Urol* 146: 463–468
60. Ball RA, Lipton SA, Dreyer EB, Richie JP, Vickers MA (1992) Entubulization repair of severed cavernous nerves in the rat resulting in return of erectile function. *J Urol* 148: 211–215
61. Te AE, Santarosa RP, Koo HP, Buttyan R, Greene LA, Kaplan SA, Olsson CA, Shabsigh R (1994) Neurotrophic factors in the rat penis. *J Urol* 152: 2167–2172
62. Dahiya R, Chui R, Perinchery G, Nakajima K, Oh BR, Lue TF (1999) Differential gene expression of growth factors in young and old rat penile tissues is associated with erectile dysfunction. *Int J Impot Res* 11: 201–206
63. Jung GW, Spencer EM, Lue TF (1998) Growth hormone enhances regeneration of nitric oxide synthase-containing penile nerves after cavernous nerve neurotomy in rats. *J Urol* 160: 1899–1904
64. Laurikainen A, Hiltunen JO, Thomas-Crusells J, Vanhatalo S, Arumae U, Airaksinen MS, Klinge E, Saarma M (2000) Neurturin is a neurotrophic factor for penile parasympathetic neurons in adult rat. *J Neurobiol* 43: 198–205
65. Sezen SF, Hoke A, Burnett AL, Snyder SH (2001) Immunophilin ligand FK506 is neuroprotective for penile innervation. *Nat Med* 7: 1073–1074
66. Podlasek CA, Zelner DJ, Jiang HB, Tang Y, Houston J, McKenna KE, McVary KT (2003) Sonic hedgehog cascade is required for penile postnatal morphogenesis, differentiation, and adult homeostasis. *Biol Reprod* 68: 423–438
67. Lee MC, El-Sakka AI, Graziottin TM, Ho HC, Lin CS, Lue TF (2002) The effect of vascular endothelial growth factor on a rat model of traumatic arteriogenic erectile dysfunction. *J Urol* 167: 761–767
68. Deng DY, Lin G, Bochinski D, Lin C, Lue TF (2003) Effect of prostaglandin E1 on angiogenesis and neural growth. *J Urol* 169: 313 (abstract 1219)
69. Burnett AL, Becker RE (2004) Immunophilin ligands promote penile neurogenesis and erection recovery after cavernous nerve injury. *J Urol* 171: 495–500
70. Bochinski DJ, Hsieh PS, Chen KC, Lin GT, Yeh CH, Lin CS, Nunes L, Deng D, Lue TF (2003) The effect of vascular endothelial growth factor and brain derived neurotrophic factor on cavernous nerve regeneration in a nerve-crush rat model. *J Urol* 169: 306 (abstract 1190)
71. Bochinski DJ, Hsieh PS, Chen KC, Yeh CH, Nunes L, Lin GT, Lin CS, Spencer M, Deng D, Lue TF (2003) The effect of insulin-like growth factor-1 and insulin-like growth factor binding protein-3 in cavernous nerve cryoablation. *J Urol* 169: 313 (abstract 1215)
72. Bakircioglu ME, Lin CS, Fan P, Sievert KD, Kan YW, Lue TF (2001) The effect of adeno-associated virus mediated brain derived neurotrophic factor in an animal model of neurogenic impotence. *J Urol* 165: 2103–2109
73. Gholami SS, Rogers R, Chang J, Ho HC, Grazziottin T, Lin CS, Lue TF (2003) The effect of vascular endothelial growth factor and adeno-associated virus mediated brain derived neurotrophic factor on neurogenic and vasculogenic erectile dysfunction induced by hyperlipidemia. *J Urol* 169: 1577–1581
74. Kim JH, Bennett NE, Sasaki K, Yoshimura N, Wolfe DP, Goins WF, Nelson JB, DeGroat WC, Glorioso JC, Chancellor MB (2003) Neurotrophic factor gene therapy: potential cure for post radical prostatectomy erectile dysfunction. *J Urol* 169: 303 (abstract 1179)
75. Magee TR, Davila HH, Ferrini MG, Rajfer J, Gonzalez-Cadavid NF (2003) Gene transfer of antisense pin (protein inhibitor of NOS CDNA) ameliorates aging-related erectile dysfunction in the rat. *J Urol* 169: 305 (abstract 1184)
76. Graziottin TM, Resplande J, Nunes L, Rogers R, Gholami S, Lue T (2002) Long-term survival of autotransplanted major pelvic ganglion in the corpus cavernosum of adult rats. *J Urol* 168: 362–366
77. Kim JH, Bennett N, Yoshimura N, Chermansky C, Kwon DD, Tarin T, Nelson JB, DeGroat WC, Qu-Petersen Z, Huard J et al. (2003) Neurorecovery and improved erectile function using muscle

derived stem cells (MDSC) in a model of post-radical prostatectomy erectile dysfunction. *J Urol* 169: 323 (abstract 1256)
78 May F, Weidner N, Mrva T, Caspers C, Matiasek K, Ganesbacher B, Hartung R (2003) Nerve guides seeded with Schwann cells are very effective for repair of ablated cavernosal nerves in rats. *J Urol* 169: 304 (abstract 1181)
79 Connolly SS, Yoo JJ, Soker S, McDougal WS, Atala A (2003) Autonomic nerve regeneration for erectile function. *J Urol* 169: 324 (abstract 1257)
80 Steiner JP, Dawson TM, Fotuhi M, Glatt CE, Snowman AM, Cohen N, Snyder SH (1992) High brain densities of the immunophilin FKBP colocalized with calcineurin. *Nature* 358: 584–587
81 Steiner JP, Dawson TM, Fotuhi M, Snyder SH (1996) Immunophilin regulation of neurotransmitter release. *Mol Med* 2: 325–333
82 Snyder SH, Lai MM, Burnett PE (1998) Immunophilins in the nervous system. *Neuron* 21: 283–294
83 Gold BG, Katoh K, Storm-Dickerson T (1995) The immunosuppressant FK506 increases the rate of axonal regeneration in rat sciatic nerve. *J Neurosci* 15: 7509–7516
84 Lee M, Doolabh VB, Mackinnon SE, Jost S (2000) FK506 promotes functional recovery in crushed rat sciatic nerve. *Muscle Nerve* 23: 633–640
85 Lyons WE, Steiner JP, Snyder SH, Dawson TM (1995) Neuronal regeneration enhances the expression of the immunophilin FKBP-12. *J Neurosci* 15: 2985–2994
86 Freeman EE, Grosskreutz CL (2000) The effects of FK506 on retinal ganglion cells after optic nerve crush. *Invest Ophthalmol Vis Sci* 41: 1111–1115
87 Kihara M, Kamijo M, Nakasaka Y, Mitsui Y, Takahashi M, Schmelzer JD (2001) A small dose of the immunosuppressive agent FK506 (tacrolimus) protects peripheral nerve from ischemic fiber degeneration. *Muscle Nerve* 24: 1601–1606
88 Sezen SF, Blackshaw S, Steiner JP, Burnett AL (2002) FK506 binding protein 12 is expressed in rat penile innervation and upregulated after cavernous nerve injury. *Int J Impot Res* 14: 506–512

Neurologic erectile dysfunction

Michael Müntener[1] and Brigitte Schurch[2]

[1] Department of Urology, University Hospital, Frauenklinikstr. 10, 8091 Zürich, Switzerland
[2] Spinal Cord Injury Center, University Hospital Balgrist, Forchstrasse 340, 8008 Zürich, Switzerland

Introduction

Because erection is a neurovascular event, any disease or dysfunction affecting the brain, spinal cord, cavernosus and pudendal nerves, or receptors in the terminal arterioles and cavernosus muscles can induce erectile dysfunction (ED). The medial preoptic area with the paraventricular nucleus has been regarded as an important integration centre for sexual drive and penile erection in animal studies [1]. Pathological processes affecting these regions, such as Parkinson's disease or stroke, are often associated with ED. Parkinsonism's effect may be caused by the imbalance of dopaminergic pathways. In patients with spinal cord injury, the degree of erectile function that persists depends largely on the nature, location and extent of the spinal lesion. Other disorders at the spinal level (e.g., spina bifida, disc herniation, syringomyelia, tumor and multiple sclerosis) may affect the afferent and efferent neural pathways in a similar manner.

This Chapter will concentrate on the incidence and pathophysiological aspects, diagnostic tools and treatment options of three main neurological diseases that affect sexual function: spinal cord injury (SCI), multiple sclerosis (MS) and Parkinson's disease (PD). Most issues discussed herein will apply to neurogenic erectile dysfunction of spinal or supraspinal origin in general.

Physiology of male sexual function

Sexual stimulation of the human males results in a series of psychological, neuronal, vascular, and local genital changes. The psychosexual response cycle in human consists of four phases: excitement, plateau, orgasm, and resolution. Another classification has characterized the dynamic changes of the penis, more exact of the corpora cavernosa during the sexual cycle. In this classification each of the psychosexual phases is divided into two interrelated events as follows: excitement into latency and tumescence, plateau into erection and rigidity, orgasm into emission and ejaculation, and resolution into detumescence and refractoriness. Male erection is an integrated component of a complex series of physiological processes that encompass male sexual behavior. Erection is a vascular event associated with tumescence of the cavernous bodies that relies upon integration of neural and humoral mechanisms at various levels of the neuraxis [2, 3]. It is relatively unique among visceral functions in that it requires central neural input for proper function. Sexual function requires participation of autonomic and somatic nerves and integration of numerous spinal and supraspinal sites in the central nervous system. A spinal generator system exists for the control of erection that can be activated by the pudendal, pelvic and possibly the hypogastric nerve afferents. The spinal system is under inhibitory and excitatory modulatory control from the brain. Ascending pathways modulate activity or level setting of the supraspinal sites as well. These ascending and descending pathways modulate the spinal generator system at the level of the spinal interneurons found in the intermediolateral column and dorsal interneurons. The central anatomical structures or networks involved in the mediation of erection include the prefrontal cortex, hippocampus, amygdala, hypothalamus, midbrain, pons and medulla. Supraspinal sites that project directly to the spinal cord generator include the paraventricular nucleus, the locus coeruleus, nucleus paragigantocellularis, parapyramidal reticular formation, raphe magnus, raphe pallidus, A5-adrenergic cell group and Barrington's nucleus. Among the major neurotransmitters involved in the supraspinal control of erection are dopamine, norepinephrine, serotonin, oxytocin, glutamate, and other neuropeptides and hormones [4]. Male erection is only a single component of this complex behavior and occurs in response to tactile, visual, and imaginative stimuli. An appreciation of the organization of neurotransmitters within neural pathways to and from the penis offers insight into the mechanisms responsible for penile erection and processes causing ED.

The sacral spinal cord is essential for penile erection. Tactile stimulation of the penis or supraspinal stimuli received and originating within the brain can elicit an erection. Excitatory and inhibitory mechanisms within the spinal cord coordinate and integrate these neural inputs. Tactile stimulation of the penis, under appropriate conditions, results in an erection even after a suprasacral SCI. This has been used as evidence that the neural network responsible for erection requires a spinal circuit. Suprasacral transection prevents supraspinal inhibition exemplified by the observation that innocuous tactile input can trigger a reflexogenic erection, albeit of short duration [5]. The major neurologic stimulus for ejaculation is the sympathetic nervous system. In men, this leads to peristaltic waves in the smooth muscle of the ampulla, seminal vesicles, and prostate. The sympathetic nervous system also enables bladder neck closure to prevent retrograde ejaculation [2]. Rhythmic contractions of the bulbocavernosus, ischiocavernosus, and pelvic musculature, which are under control of the somatic nervous system, then result in antegrade ejection of semen to the urethral meatus.

Pathophysiology

Spinal cord injury and sexual dysfunction

Male sexual dysfunction after SCI involves ED, disorders of ejaculation, orgasmic dysfunction and failure of detumescence (priaprism). The incidence of ED after SCI approximates 75%. Patients with incomplete SCI are more likely to achieve erections sufficient for sexual intercourse than complete patients. Also, patients with upper motorneuron lesion (UMN) lesions (above the sacral spinal cord) have more frequent erections than those with lower motoneuron lesion (LMN) lesions (at the level or below the sacral spinal cord). In one study 92% of SCI patients with UMN lesions were able to achieve some degree of reflexogenic erection [6]. These reflex erections may be of short duration and insufficient for intercourse. Erections in SCI patients are classified into two categories: psychogenic and reflexogenic. Reflex erections are parasympathetically mediated through a local sacral (S2–S4) reflex arc. They tend to occur in patients with UMN lesions. LMN lesions can interrupt the sacral reflex arc and prevent erections in response to tactile stimulation. In comparison, psychogenic erections are thought to be mediated by sympathetic pathways that exit the spinal cord at T10–T12. The role of the sympathetic nervous system in eliciting erection has been controversially discussed. Root and Bard already demonstrated that cats could maintain penile responses in the presence of a receptive female despite complete destruction of the sacral pathways [7]. This finding was counterbalanced by other negative findings, whereby electrical stimulation of the nervi erigentes failed to elicit penile responses or elicited only a few responses following stimulation of the distal portion of the nerves [8, 9]. However, observations from spinal cord injured men who lost reflex activity from the genital area showed that many of them still had erections elicited by supraspinal stimuli. Since innervation of the male reproductive system suggests that two neuronal pathways supply the genitals (sacral and thoracolumbar), it is presumed that the second thoracolumbar pathway should compensate for the loss of the first sacral pathways in case of low spinal cord injury [3, 10–17]. Transection and stimulation experiments of the hypogastric nerve (HGN) in rats [17, 18] and cats [7] underline these clinical findings. However, if in fact the HGN mediates psychogenic erection in men with spinal cord injury, this does not necessarily mean that HGN mediates psychogenic erection in normal male subjects. Considerable neuroplasticity, occurring after some time, may have enabled or potentiated an erectile function for the HGN which was absent or minimal before trauma [8, 17]. This means for clinical practice that patients with LMN lesions have not to be considered as irreversibly impotent and might profit for adequate treatment.

In summary: male spinal cord injured patients present with a high incidence of erectile dysfunction. The ability of achieving reflexogenic or psychogenic erection depends on the level of the spinal cord lesion, whereas the extent of the lesion determines the quality of the erection achieved.

Sexual dysfunction and multiple sclerosis

Sexual dysfunction is a significant but often underestimated symptom of MS, since it severely affects the quality of life of MS patients [19]. Indeed, it is not surprising that a disorder like MS, which produces demyelinating lesions all over the white matter of the brain and the spinal cord, can cause sexual disturbances, frequently, though not invariably, associated with sphincteric dysfunction [20, 21]. During the course of MS, symptoms of sexual dysfunction become increasingly distressing, affecting 10–20% of patients, after 10 years of disease duration [22, 23]. In a five-year follow-up study comprising 49 patients with definite MS and more than 10 years duration, Stenager et al. demonstrated a significant increase over time of the risk of sexual dysfunction [24]. The percentage of patients with changed sexuality rose significantly from 43–71% in the observation period and the change was independent from age at the onset of symptoms, duration of disease, gender and disability. In their two year follow-up, Zorzon et al. did not report an increase in the number of patients with at least new symptoms of ED, but an increase of extent and number of symptoms, particularly in men who reported changes of sexual function significantly more frequently than women [25]. Insufficient length of follow-up, low relapse rate, absence of change in disability, increasing use of interferon therapy might explain the lack of increase in the number of patients with sexual dysfunction in Zorzon's study. In most of the studies on MS patients an association between sexual dysfunction and sphincteric dysfunction has been found [21, 25–31].

In summary: the relationship between sexual dysfunction and neurological impairment, disability, age, disease's course, duration of disease and fatigue in MS patients is with exception of Zorzon's study generally accepted [21, 25–29, 31, 32]. The positive correlation between bladder dysfunction and sexual dysfunction suggests that bladder and sexual function share the same segments of autonomic control.

Sexual dysfunction and Parkinson's disease

Treatment of PD is primarily aiming at improving motor function. However, especially in advanced stages, PD is often complicated by additional problems such as treatment related complications, falls, and depression, which may have much greater impact on the Parkinson's patients quality of life, than the cardinal features of PD. Sexual dysfunction has been observed in men with PD and ED seems to be correlated with disease severity and depressive symptoms [33–35]. In these patients the higher prevalence of ED can be related to motor symptoms and the use of anti-Parkinsonian drugs [36], or often can be a consequence of reduced dopaminergic tone and high incidence of depression (often prior to the onset motor symptoms) [37, 38], such as in patients treated with anti-dopaminergic drugs (such as reserpine or alpha-methyldopa) that reduce the functional efficiency of dopaminergic terminals [39–42], or antipsychotic drugs which antagonize dopamine receptors in the CNS.

Depression is a common disturbance reported in patients with PD, with a frequency approaching 40% [43, 44].

In their study on sexuality in young patients with PD as compared to healthy controls, Jacobs et al. found that depression in Parkinson patients was more relevant to the subjective dissatisfaction rather than physiological dysfunction [45].

For physiopathological explanation, several observations suggest the involvement of the central dopaminergic system in the overall etiology of erectile dysfunction [33, 34, 40]. In the last 10 years, however, attention has increasingly focused on the mesolimbic system, which is now recognized as the major reward pathway in the CNS. The last pathway coincides with the mesolimbic dopaminergic system, which projects from the ventral segmental areas (a mesencephalic region located in the vicinity of the pars compacta of the substantia nigra) to the nucleus accumbes and other limbic areas [46]. The mesolimbic dopaminergic pathway responds to natural or artificial rewards by generating the sensation of pleasure, by encoding the memory of pleasurable cues and by promoting motor and behavioral response that direct the individual towards the source of pleasure.

In summary: most studies on Parkinson patients suggest that depression is a key factor in the appearance of sexual dysfunction. Young PD patients do not differ in their sexual function from age-matched controls, but their perception of sexual functioning and general health is considerably influenced by depression. In more advanced stage, disease severity and intake of anti-Parkinsonian drugs certainly contribute to ED and should be taken into account in the evaluation of ED by PD patients.

Diagnosis

Neurogenic ED is usually diagnosed in impotent patients suffering from a concomitant neurologic condition (i.e., SCI, MS, PD, etc.) that is able to cause impairment of erectile function. If no such disease is present neurologic origin of the problem is unlikely and neurologic ED should only be diagnosed if other possible reasons have been excluded.

Because the treatment of neurologic ED as well as the treatment of ED in general is goal-orientated and usually symptomatic, the clinical value of most diagnostic tests is limited. Diagnostic routine in neurologic ED, however, has two mayor goals: one is to exclude relevant concomitant medical conditions requiring treatment and the second is to assess the needs and expectations of each patient in regard to the management of his ED. Several special tests can additionally be helpful to tailor the therapy optimally to the individual patient. Their respective diagnostic value is discussed below.

Certainly it is reasonable not only to individualise the therapy but also the diagnostic workup in patients with neurologic ED. In an otherwise healthy young man with a traumatic spinal cord injury and normal erectile function prior to the accident basic evaluation is usually simpler than in an elderly man with Parkinson's disease, gradual loss of erectile function and severe concomitant medical conditions.

Medical History

Taking the patient's history should aim at detecting treatable cofactors or conditions associated with ED (i.e., cardiovascular disease, diabetes, drug abuse etc.).

A history of diabetes mellitus, MS, PD or other chronic neurologic disorders as well as a history of SCI, pelvic surgery, trauma and irradiation may indicate neurogenic ED.

A thorough sexual history is the cornerstone of diagnostic procedures also in neurologic ED. Patients should be encouraged to bring along their partner because the problem affects the couple and information provided by the patient's partner can be very helpful.

Sexual history should reveal relevant psychological cofactors. Questions to be answered are whether the onset of the problem was sudden or gradual, whether there is some form of trigger, whether the libido is normal and whether there is concomitant ejaculatory or orgasmic disorder. And patients should be asked whether they can still achieve some degree of an erection and if so of what quality these erections are [47]. Important information is, which stimulus (psychogenic or reflexogenic) produces the best erectile response. Standardised questionnaires like the International Index of Erectile Function (IIEF) [14] are routinely used to assess erectile function in clinical studies. Also in everyday clinical practice they can be helpful to protocol a patient's follow-up. However, no questionnaire can replace careful history taking [48].

Physical examination

Every patient with ED should undergo a routine physical examination. If a neurogenic cause is suspected, a full neurologic evaluation should be performed. Spinal cord injured patients should be examined according to the ASIA scores [49]. Special attention has to be given to eventual sensory impairment of the sacral dermatoms, the loss or preservation of voluntary contraction of the external anal sphincter as well as the bulbocavernous reflex (BCR). BCR is a response that is recorded from the bulbocavernosus muscle after stimulation of the pudendal nerve, usually on the penis or clitoris. The BCR reveals the integrity of the spinal segments of S2–3 and their afferent and efferent connections in the urogenital region including the sensory and motor fibres of the pudendal nerve. BCR is considered pathological if absent or by latency above

45 msec in electrophysiologic evaluation [50]. Loss of BCR correlates strongly with a loss of reflexogenic erections in spinal cord injured patients [51].

Laboratory tests

Recommended blood tests are basically no different from routine workup in non-neurogenic ED. They should rule out hypogonadism, diabetes mellitus, hypercholesterolemia as well as hepatic or renal failure. In an otherwise healthy young man with ED due to SCI, however, additional lab tests can probably be omitted.

Special diagnostic tests

Unfortunately today there is no reliable test available to assess the autonomic function of the corpora cavernosa. The standard tests of genital neurologic function rely on somatic nervous system diagnostic tests or evaluate the autonomic function indirectly [52]. The most often used of these tests as well as their respective role in the diagnosis of neurogenic erectile dysfunction are described below.

Pharmacotesting
Pharmacotesting can be used to rule out severe concomitant arteriogenic or veno-occlusive dysfunction and consists of intracavernous injection of a vasoactive substance (e.g., Prostaglandin E1) and the subsequent rating of the induced erection [53]. If readily available pharmacotesting can be combined with assessment of hemodynamic characteristics by Doppler ultrasonography.

Audio-visual sexual stimulation (AVSS)/vibrator stimulation
With AVSS patients watch an erotic movie or read erotic literature. If patients are asked not to masturbate AVSS can be used to evaluate psychogenic erections [15]. Correspondingly the quality of reflexogenic erections can be assessed by vibrator stimulation of the glans penis. Psychogenic and reflexogenic erections can be evaluated this way in spinal cord injured patients. The results of these tests have been shown to be very consistent with the patient's self reports [51], therefore a careful sexual history yields a similar diagnostic accuracy.

Nocturnal penile tumescence and rigidity testing (NPTR)
In NPTR erectile activity, which physiologically occurs in healthy males during rapid eye movement stages of sleep, is measured in the sleeping patient. Ideally this test is performed in a sleep laboratory on two to three consecutive nights.

However, nocturnal and sexually induced erections cannot be fully equated and it is likely that they differ primarily in neurological not vascular mechanisms. It is also known that there are patients with neurologic erectile dysfunction who have normal NPTR [54]. A normal NPTR rules out major vascular dysfunction, whereas an abnormal NPTR is of very little diagnostic value. The role of NPTR remains controversial [55].

Pudendal somatosensory evoked potentials (P-SSEP)
P-SSEP are elicited by electrical stimulation on the penis whereas the central responses and their respective latencies are registered by a scalp electrode. P-SSEP have been used to assess the afferent limb of the sacral reflex arc as well as to provide information about the integrity of the sensory spinobulbar tract [56]. Thus P-SSEP indirectly tests the connection of the supraspinal and the sacral erection centers. In spinal cord injured patients normal and pathologic P-SSEP correlate very well with preserved and absent sensibility in the sacral dermatoms [51], respectively.

Sympathetic skin response (SSR)
Recording of perineal SSR is an electrophysiologic test where sudomotor activity is measured on the perineal skin in response to an electrical stimulus to the median nerve at the wrist [57]. Hence a positive perineal SSR proves efferent sympathetic innervation of the genital region. In this sense SSR has been found helpful in the diagnosis of neurologic ED of various etiologies [58]. In spinal cord injured patients positive SSR correlate very well with the preservation of psychogenic erections [18, 51].

Urodynamics
In many patients with neurogenic ED there is some form of concomitant bladder dysfunction [52]. In neurogenic bladder dysfunction urodynamic studies are routinely used to specify the type of the detrusor disorder. It is known that findings from these studies have a certain predictive value in regard to the erectile function of the respective patient [59]. Detrusor hyperreflexia and detrusor areflexia are closely associated with preservation and loss of reflexogenic erections, respectively [51].

Therapy

Counseling is a cornerstone of the management also in neurogenic erectile dysfunction. Assessing the individual patient's needs and expectations and discussing the therapeutic approach in general as well as the expected benefits of a specific therapy are crucial for long-term patient's compliance and a successful treatment. As in ED therapy in general, there are hardly causative treatment options in neurogenic ED and the therapy will therefore treat the symptom rather than the cause of the problem. Although this connection is obvious for physicians, it needs to be explained to most patients, and those who understand it, will be more satisfied with the treatment of their ED.

If relevant non-urological comorbidities are detected or suspected in the diagnostic workup the patient should be referred to a specialist to further evaluate and treat the respective condition. Special attention has to be given to psychological changes that frequently accompany neurogenic erectile dysfunction of different etiologies. In PD for instance, the frequency of depression is reported to be about 40% [43]. Also spinal cord injured patients commonly suffer at least temporarily from reactive depressive episodes that can additionally impair their erectile function and many spinal cord injured patients may benefit from a psychotherapeutic workup [60]. Psychosexual therapy can therefore be recommended generously in patients with neurologic ED. It is one of the very few cause-oriented treatment options in the management of ED and it can be combined with pharmacological therapy.

General considerations

Recommendations for paraplegic patients, which possibly apply also to patients with neurogenic ED of other etiologies are precoital emptying of the urinary bladder as well as application of warm towels to the penis [6]. Additionally penile vibrator stimulation can be used to improve erections and application of a penile constriction device (i.e., ring or band) can help to enhance and maintain penile rigidity [61].

Sexual arousal can provoke autonomic dysreflexia in spinal cord injured patients with a lesion above T6. This condition is characterised by an increase in blood pressure causing a pounding headache and it can be accompanied by flushing, sweating and cardiac arrhythmias [62]. Patients susceptible for autonomic dysreflexia should be instructed to empty their bladder and rectum before sexual stimulation as well as to stop the stimulation immediately when respective symptoms occur. Additionally the prophylactic use of nifedipine has been recommended [63].

Sildenafil (Viagra®) in the treatment of neurogenic erectile dysfunction

Independent of the underlying disease, sildenafil is a first line choice in the treatment of neurologic erectile dysfunction. Since its tolerability and efficacy have been proven in many controlled and uncontrolled clinical trials, to date sildenafil is the best-studied oral medical therapy for this condition. Clinical experience with sildenafil treatment in patients with SCI, MS and PD are discussed below, however, sildenafil is also effective for neurogenic ED of less frequent etiologies [64].

Although sildenafil may as well have central effects [52, 65], its main mode of action is peripheral and permits the enhancement of both reflexogenic and psychogenic erections. The more erectile function is preserved in a patient the more likely is he to respond to an oral pharmacological treatment with sildenafil [66]. Nevertheless sildenafil can also improve the erectile function of patients with little or no residual function [67].

Spinal cord injury

The efficacy and tolerability of sildenafil in the treatment of spinal cord injured patients has been investigated in two randomised, placebo-controlled trials [68–71] as well as four case series [66, 72–74]. The sample size in these studies ranged from 17 [74] to 178 [70, 71] and in most of the studies patients with cervical to lumbar lesion levels as well as both incomplete and complete lesions were enrolled. The results of these studies have been summarised and discussed in detail by Derry and co-workers [67].

In general, sildenafil was significantly more effective than placebo in the treatment of erectile dysfunction of spinal cord injured men. The proportion of patients who reported an improvement of their erections under sildenafil was up to 94% and up to 72% of intercourse attempts were successful in treated patients.

Adverse events were usually mild and did not differ from those seen in the general population. Particularly autonomic dysreflexia and priapism have not been reported in any of the studies. The proportion of patients who discontinued the treatment due to adverse events was about 5%. Treatment with sildenafil has also been shown to significantly improve the quality of life and the overall satisfaction with sexual life in spinal cord injured patients [71].

Multiple sclerosis

The effect of sildenafil in men with MS and ED was examined in a double-blind placebo-controlled trial involving 217 patients [75, 76] as well as in a small case series involving seven patients [73].

Treatment with sildenafil improved erections significantly better than placebo with 90% and 24% of patients reporting better erections, respectively, after 12 weeks of treatment. The drug was well tolerated and only the well-known, mild adverse events were reported. Additionally significant improvements in general areas of quality of life were demonstrated.

Parkinson's disease

One double-blind placebo-controlled trial [77] and two prospective open-label studies [78, 79] evaluated the efficacy and safety of oral sildenafil in patients with ED and PD. All of these trials had small sample sizes ranging from 10–33 patients. However, sildenafil improved the erectile function of treated patients in all studies. Raffaele and co-workers who investigated the efficacy and safety of sildenafil in the treatment of 33 depressed patients with idiopathic PD found that sildenafil indirectly also improved depressive symptoms [79].

Summary

In summary sildenafil is an effective oral medical therapy for neurologic ED. It is safe, well tolerated and it improves erectile function as well as overall satisfaction with sexual life and the quality of life in treated patients.

Alternative oral therapies

Newer phosphodiesterase 5 inhibitors like vardenafil and tadalafil have generally been proven effective in the treatment of ED [80, 81]. Obviously the clinical experience with these agents is not as large as with sildenafil but their efficacy and safety profile can be expected to be very similar to sildenafil. Like sildenafil, vardenafil and tadalafil can be regarded as first line treatment options for neurogenic ED.

An additional oral therapy is the centrally acting dopamine receptor agonist apomorphine. Although the onset of action is reportedly faster than with sildenafil, apomorphine is generally less effective [82]. In a recent study the efficacy of apomorphine for the treatment of erectile dysfunction in spinal cord injured men was disappointing [83].

Second line treatment options

As in non-neurogenic ED second line medical therapy consists of intraurethral or intracavernosal injection of erectogenic agents (e.g., Prostaglandin E1). These therapies have been shown to be effective [84, 85] and in patients with neurogenic ED usually lower dosages are needed than in patients with vasculogenic disease [52]. Disadvantages are an increased risk of priapism and the need of good manual dexterity for correct application of the drug.

A non-invasive treatment option is the use of vacuum constriction devices. Although proven to be effective in patients with neurogenic ED [86], the handling of these devices can be troublesome and since the introduction of efficacious oral therapies they have lost more and more general acceptance.

Third line treatment options

The last resort in the therapy of ED is the implantation of a penile prosthesis. This invasive and irreversible treatment carries a significant risk of infection, which particularly can be a problem in spinal cord injured men [87].

References

1 Sachs BD, Meisel RJ (1988) The physiology of male sexual behavior. In: E Knobil, JD Neil, LL Ewing (eds) *The physiology of reproduction*. Raven Press, New York, 1393–1423
2 Andersson KE, Wagner G (1995) Physiology of penile erection. *Physiol Rev* 75: 191
3 De Groat WC, Steers WD (1988) Neuroanatomy and neurophysiology of penile erection. In: EA Tanagho, TF Lue, RD McClure (eds): *Contemporary management of impotence and infertility*. Williams and Wilkins, Baltimore, 3–27
4 Giuliano F, Rampin O (2000) Central neural regulation of penile erection. *Neurosci Biobehav Rev* 24: 517

5 Bell C (1972) Autonomic nervous control of reproduction: circulatory and other factors. *Pharmacol Rev* 24: 657
6 Comarr AE (1970) Sexual function among patients with spinal cord injury. *Urol Int* 25: 134
7 Root WS, Bard P (1947) The mediation of feline erection through sympathetic pathways with some remarks on sexual behavior after deafferentation of the genitalia. *Am J Physiol* 151: 80
8 Cruz MR, Liu YC, Manzo J, Pacheco P, Sachs BD (1999) Peripheral nerves mediating penile erection in the rat. *J Auton Nerv Syst* 76: 15
9 Semens JH, Longworthy OR (1938) Observations on the neurophysiology of sexual function in the male car. *J Urol* 40: 836
10 Chapelle PA, Durand J, Lacert P (1980) Penile erection following complete spinal cord injury in man. *Br J Urol* 52: 216
11 Jänig W, McLachlan EM (1987) Organization of lumbar spinal outflow to distal colon and pelvic organs. Physiol Rev 67: 1332
12 Bors E, Comarr AE (1960) Neurophysiological disturbance of sexual function with special reference to 529 subjects with spinal cord injury. *Urol Surv* 10: 191
13 Sachs BD (1995) Placing erection in context: the reflexogenic-psychogenic dichotomy reconsidered. *Neurosci Biobehav Rev* 19: 211
14 Rosen RC, Riley A, Wagner G, Osterloh IH, Kirkpatrick J, Mishra A (1997) The international index of erectile function (IIEF): a multidimensional scale for assessment of erectile dysfunction. *Urology* 49: 822
15 Courtois FJ, Goulet MC, Charvier KF, Leriche A (1999) Posttraumatic erectile potential of spinal cord injured men: how physiologic recordings supplement subjective reports. *Arch Phys Med Rehabil* 80: 1268
16 Courtois FJ, Macdougall JC, Sachs BD (1993) Erectile mechanism in paraplegia. *Physiol Behav* 53: 721
17 Dail WG, Walton G, Olmsted MP (1989) Penile erection in the rat: stimulation of the hypogastric nerve elicits increases in penile pressure after chronic interruption of the sacral parasympathetic outflow. *J Auton Nerv Syst* 28: 251
18 Courtois FJ, Charvier KF, Leriche A, Raymond DP (1993) Sexual function in spinal cord injury men. I. Assessing sexual capability. *Paraplegia* 31: 771
19 Dupont S (1995) Multiple sclerosis and sexual functioning. *Clin Rehab* 9: 135
20 Stenager E, Stenager EN, Jensen K (1992) Sexual aspects of multiple sclerosis. *Semin Neurol* 12: 120
21 Bakke A, Myhr KM, Gronning M, Nyland H (1996) Bladder, bowel and sexual dysfunction in patients with multiple sclerosis – a cohort study. *Scand J Urol Nephrol Suppl* 179: 61
22 Zorzon M, Zivadinov R, Bosco A, Bragadin LM, Moretti R, Bonfigli L, Morassi P, Iona LG, Cazzato G (1999) Sexual dysfunction in multiple sclerosis: a case-control study. I. Frequency and comparison of groups. *Mult Scler* 5: 418
23 Stenager E, Stenager EN, Jensen K et al. (1990) Multiple sclerosis: sexual dysfunction. *J Sex Educ Ther* 16: 262–269
24 Stenager E, Stenager EN, Jensen K (1996) Sexual function in multiple sclerosis: a 5 year follow-up study. *Ital J Neurol Sci* 17: 227
25 Zorzon M, Zivadinov R, Monti Bragadin L, Moretti R, De Masi R, Nasuelli D, Cazzato G (2001) Sexual dysfunction in multiple sclerosis: a 2-year follow-up study. *J Neurol Sci* 187: 1
26 Minderhoud JM, Leemhuis JG, Kremer J, Laban E, Smits PM (1984) Sexual disturbances arising from multiple sclerosis. *Acta Neurol Scand* 70: 299
27 Valleroy ML, Kraft GH (1984) Sexual dysfunction in multiple sclerosis. *Arch Phys Med Rehabil* 65: 125
28 Hulter BM, Lundberg PO (1995) Sexual function in women with advanced multiple sclerosis. *J Neurol Neurosurg Psychiatry* 59: 83
29 Mattson D, Petrie M, Srivastava DK, McDermott M (1995) Multiple sclerosis. Sexual dysfunction and its response to medications. *Arch Neurol* 52: 862
30 Vas J (1969) Sexual Impotence and some autonomic disturbances in men with multiple sclerosis. *Acta Neurol Scand* 45: 166
31 Zivadinov R, Zorzon M, Bosco A, Bragadin LM, Moretti R, Bonfigli L, Iona LG, Cazzato G (1999) Sexual dysfunction in multiple sclerosis: II. Correlation analysis. *Mult Scler* 5: 428
32 Lilius HG, Valtonen EJ, Wikstrom J (1976) Sexual problems in patients suffering from multiple sclerosis. *Scand J Chron Nephrol Suppl* 179: 61

33 Brown RG, Jahanshahi M, Quinn N, Marsden CD (1990) Sexual function in patients with Parkinson's disease and their partners. *J Neurol Neurosurg Psychiatry* 53: 480
34 Cleeves L, Findley LJ (1987) Bromocriptine induced impotence in Parkinson's disease. *Br Med J (Clin Res Ed)* 295: 367
35 Helfried J, Vieregge A, Vieregge P (2000) Sexuality in young patients with Parkinson's disease: A population based comparison with healthy controls. *J Neurol Neurosurg Psychiatry* 69: 550
36 Lipe H, Longstreth WT Jr, Bird TD, Linde M (1990) Sexual function in married men with Parkinson's disease compared to married men with arthritis. *Neurology* 40: 1347
37 Commings J (1985) Psychosomatic aspects of movements disorders. *Adv Psychosomatic Med* 13: 111
38 Mayeux R (1990) Parkinson's disease. *J Clin Psychiatry* 51: 20
39 Randrup A, Munkvad I, Fog R (1975) Mania depression and brain dopamine. In: WB Essman, L Valzelli (eds): *Current Development in Psychopharmacology*. Spectrum Publications, New York, 207–229
40 Goodwin FK, Sack RL (1974) Central dopamine function in affective illness: Evidence from precursors, enzyme inhibitors, and studies of central dopamine turnover. In: E Usdin (ed): *Neuropharmacology of Monoamines and their Regulatory Enzymes*. Raven, New York, 261–277
41 McKinney WT, Kane FJ (1967) Depression with the use of alpha-methyldopa. *Am J Psychiatry* 124: 118
42 Willner P (1983) Dopamine and depression: A review of recent evidence. Part II: Theoretical approaches. *Brain Res* 287(3): 225
43 Cummings JL (1992) Depression and Parkinson's disease: a review. *Am J Psychiatry* 149: 443
44 Gotham AM, Brown RG, Marsden CD (1986) Depression in Parkinson's disease: a quantitative and qualitative analysis. *J Neurol Neurosurg Psychiatry* 49: 381
45 Jacobs H, Vieregge A, Vieregge P (2000) Sexuality in young patients with Parkinson's disease: a population based comparison with healthy controls. *J Neurol Neurosurg Psychiatry* 69: 550
46 Koob GF (1998) Drug abuse and alcoholism. Overview. *Adv Pharmacol* 42: 969
47 Muntener M, Suter S, Praz V, Hauri D (2003) Erectile dysfunction: reasonable diagnostics and treatment in general practice. *Schweiz Rundsch Med Prax* 92: 179
48 Blander DS, Sanchez-Ortiz RF, Broderick GA (1999) Sex inventories: can questionnaires replace erectile dysfunction testing? *Urology* 54: 719
49 Maynard FM Jr, Bracken MB, Creasey G, Ditunno JF Jr, Donovan WH, Ducker TB, Garber SL, Marino RJ, Stover SL, Tator CH et al. (1997) International Standards for Neurological and Functional Classification of Spinal Cord Injury. American Spinal Injury Association. *Spinal Cord* 35: 266
50 Ertekin C, Akyurekli O, Gurses AN, Turgut H (1985) The value of somatosensory-evoked potentials and bulbocavernosus reflex in patients with impotence. *Acta Neurol Scand* 71: 48
51 Schmid DM, Curt A, Hauri D, Schurch B (2003) Clinical value of combined electrophysiological and urodynamic recordings to assess sexual disorders in spinal cord injured men. *Neurourol Urodyn* 22: 314
52 Nehra A, Moreland RB (2001) Neurologic erectile dysfunction. *Urol Clin North Am* 28: 289
53 Wyndaele JJ, de Meyer JM, de Sy WA, Claessens H (1986) Intracavernous injection of vasoactive drugs, an alternative for treating impotence in spinal cord injury patients. *Paraplegia* 24: 271
54 Broderick GA (1998) Evidence based assessment of erectile dysfunction. *Int J Impot Res* 10 (Suppl 2): S64
55 Morales A, Condra M, Reid K (1990) The role of nocturnal penile tumescence monitoring in the diagnosis of impotence: a review. *J Urol* 143: 441
56 Opsomer RJ, Guerit JM, Wese FX, Van Cangh PJ (1986) Pudendal cortical somatosensory evoked potentials. *J Urol* 135: 1216
57 Curt A, Weinhardt C, Dietz V (1996) Significance of sympathetic skin response in the assessment of autonomic failure in patients with spinal cord injury. *J Auton Nerv Syst* 61: 175
58 Opsomer RJ, Boccasena P, Traversa R, Rossini PM (1996) Sympathetic skin responses from the limbs and the genitalia: normative study and contribution to the evaluation of neurourological disorders. *Electroencephalogr Clin Neurophysiol* 101: 25
59 Rydin E, Lundberg PO, Brattberg A (1981) Cystometry and mictometry as tools in diagnosing neurogenic impotence. *Acta Neurol Scand* 63: 181
60 Althof SE, Levine SB (1993) Clinical approach to the sexuality of patients with spinal cord injury. *Urol Clin North Am* 20: 527

61 Francois N, Jouannet P, Maury M (1983) Genitosexual function of paraplegics. *J Urol (Paris)* 89: 159
62 Biering-Sorensen F, Sonksen J (2001) Sexual function in spinal cord lesioned men. *Spinal Cord* 39: 455
63 Steinberger RE, Ohl DA, Bennett CJ, McCabe M, Wang SC (1990) Nifedipine pretreatment for autonomic dysreflexia during electroejaculation. *Urology* 36: 228
64 Hatzichristou DG (2002) Sildenafil citrate: lessons learned from 3 years of clinical experience. *Int J Impot Res* 14 (Suppl 1): S43
65 Giuliani D, Ottani A, Ferrari F (2002) Influence of sildenafil on copulatory behaviour in sluggish or normal ejaculator male rats: a central dopamine mediated effect? *Neuropharmacology* 42: 562
66 Sanchez Ramos A, Vidal J, Jauregui ML, Barrera M, Recio C, Giner M, Toribio L, Salvador S, Sanmartin A, de la Fuente M et al. (2001) Efficacy, safety and predictive factors of therapeutic success with sildenafil for erectile dysfunction in patients with different spinal cord injuries. *Spinal Cord* 39: 637
67 Derry F, Hultling C, Seftel AD, Sipski ML (2002) Efficacy and safety of sildenafil citrate (Viagra) in men with erectile dysfunction and spinal cord injury: a review. *Urology* 60: 49
68 Derry FA, Dinsmore WW, Fraser M, Gardner BP, Glass CA, Maytom MC, Smith MD (1998) Efficacy and safety of oral sildenafil (Viagra) in men with erectile dysfunction caused by spinal cord injury. *Neurology* 51: 1629
69 Maytom MC, Derry FA, Dinsmore WW, Glass CA, Smith MD, Orr M, Osterloh IH (1999) A two-part pilot study of sildenafil (VIAGRA) in men with erectile dysfunction caused by spinal cord injury. *Spinal Cord* 37: 110
70 Giuliano F, Hultling C, El Masry WS, Smith MD, Osterloh IH, Orr M, Maytom M (1999) Randomized trial of sildenafil for the treatment of erectile dysfunction in spinal cord injury. Sildenafil Study Group. *Ann Neurol* 46: 15
71 Hultling C, Giuliano F, Quirk F, Pena B, Mishra A, Smith MD (2000) Quality of life in patients with spinal cord injury receiving Viagra (sildenafil citrate) for the treatment of erectile dysfunction. *Spinal Cord* 38: 363
72 Schmid DM, Schurch B, Hauri D (2000) Sildenafil in the treatment of sexual dysfunction in spinal cord-injured male patients. *Eur Urol* 38: 184
73 Green BG, Martin S (2000) Clinical assessment of sildenafil in the treatment of neurogenic male sexual dysfunction: After the hype. *NeuroRehabilitation* 15: 101
74 Gans WH, Zaslau S, Wheeler S, Galea G, Vapnek JM (2001) Efficacy and safety of oral sildenafil in men with erectile dysfunction and spinal cord injury. *J Spinal Cord Med* 24: 35
75 Fowler C, Miller J, Sharief M (1999) Viagra (sildenafil citrate) for the treatment of erectile dysfunction in men with multiple sclerosis. *Ann Neurol* 46: 497
76 Miller J, Fowler C, Sharief M (1999) Effect of sildenafil citrate (Viagra) on quality of life in men with erectile dysfunction and multiple sclerosis. *Ann Neurol* 46: 496
77 Hussain IF, Brady CM, Swinn MJ, Mathias CJ, Fowler CJ (2001) Treatment of erectile dysfunction with sildenafil citrate (Viagra) in parkinsonism due to Parkinson's disease or multiple system atrophy with observations on orthostatic hypotension. *J Neurol Neurosurg Psychiatry* 71: 371
78 Zesiewicz TA, Helal M, Hauser RA (2000) Sildenafil citrate (Viagra) for the treatment of erectile dysfunction in men with Parkinson's disease. *Mov Disord* 15: 305
79 Raffaele R, Vecchio I, Giammusso B, Morgia G, Brunetto MB, Rampello L (2002) Efficacy and safety of fixed-dose oral sildenafil in the treatment of sexual dysfunction in depressed patients with idiopathic Parkinson's disease. *Eur Urol* 41: 382
80 Keating GM, Scott LJ (2003) Vardenafil: a review of its use in erectile dysfunction. *Drugs* 63: 2673
81 Padma-Nathan H (2003) Efficacy and tolerability of tadalafil, a novel phosphodiesterase 5 inhibitor, in treatment of erectile dysfunction. *Am J Cardiol* 92: 19 M
82 Martinez R, Puigvert A, Pomerol JM, Rodriguez-Villalba R (2003) Clinical experience with apomorphine hydrochloride: the first 107 patients. *J Urol* 170: 2352
83 Strebel RT, Reitz A, Tenti G, Curt A, Hauri D, Schurch B (2004) Apomorphine sublingual as primary or secondary treatment for erectile dysfunction in patients with spinal cord injury. *BJU Int* 93: 100
84 Padma-Nathan H, Hellstrom WJ, Kaiser FE, Labasky RF, Lue TF, Nolten WE, Norwood PC, Peterson CA, Shabsigh R, Tam PY (1997) Treatment of men with erectile dysfunction with transurethral alprostadil. Medicated Urethral System for Erection (MUSE) Study Group. *N Engl J Med* 336: 1

85 Linet OI, Ogrinc FG (1996) Efficacy and safety of intracavernosal alprostadil in men with erectile dysfunction. The Alprostadil Study Group. *N Engl J Med* 334: 873
86 Denil J, Ohl DA, Smythe C (1996) Vacuum erection device in spinal cord injured men: patient and partner satisfaction. *Arch Phys Med Rehabil* 77: 750
87 Carson CC (1999) Complications of penile prostheses and complex implantations. In: C Carson, R Kirby, I Goldstein (eds): *Textbook of Erectile Dysfunction*. Isis Medical Media, Oxford, 435–450

Therapy of erectile dysfunction (ED) with sildenafil improves quality of life (QoL) and partnership (QoP)

Matthias J. Müller

Department of Psychiatry, University of Mainz, Untere Zahlbacher Str. 8, 55131 Mainz, Germany

Sexual dysfunction and erectile dysfunction (ED)

According to Maslow [1] and others, higher levels of individual and social functioning can only be achieved and maintained, when more basic and particularly physiological levels of functioning are satisfactory and healthy (Fig. 1). Satisfactory partnership and sexuality are core aspects of human wellbeing through the complete course of adulthood. Sexual activity is a major determinant of sexual satisfaction [2]. In men and women, sexual dysfunction as a disorder of sexual desire, arousal, or orgasm, and/or sexual pain results in

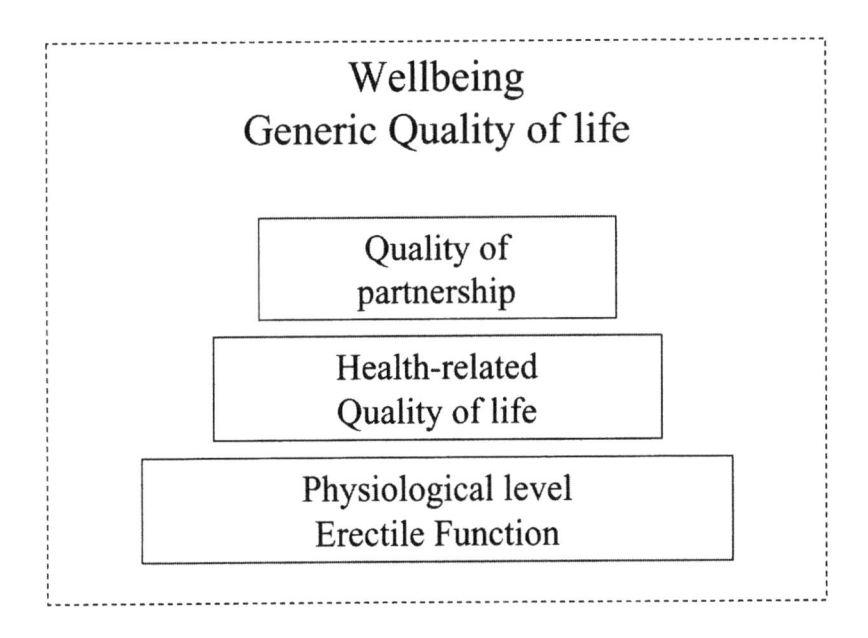

Figure 1. Erectile function and quality of life.

personal distress and reduced quality of life and interpersonal relationships [3–6]. Erectile dysfunction (ED) as the inability to attain or maintain penile erection sufficient for satisfactory sexual performance is age-associated, with estimated prevalence rates of 25–40% among men 40 years old and 60–70% among those 70 years old. ED is a common and multifactorial disease due to organic and/or psychological factors [4, 7].

ED is not a problem only for men, because the relationship between partners can also be disturbed. Therefore, adequate treatment of ED is needed and the most convenient and simplest way is oral drug therapy.

ED treatment with sildenafil

Sildenafil, a phosphodiesterase-(PDE)-5-selective inhibitor has been the drug of choice for patients with ED since it was launched in March 1998 [8]. The most frequent adverse effects documented with sildenafil usage are headache, flushes, dyspepsia, visual disturbances and nasal congestion. These adverse effects are in most cases dose-related, usually transient and mild, with a low withdrawal rate [9, 10]. Several studies performed until recently have shown that sildenafil is a safe and effective treatment of ED in patients with cardiovascular disease, who do not take nitrates or nitrate donors concomitantly.

Other oral medications for ED include apomorphine, phentolamine, yohimbine, trazodone, testosterone and new PDE5 inhibitors such as vardenafil and tadalafil. According to all available data, the concept of PDE5 inhibition has a central position in recent oral pharmacotherapy of ED. However, larger clinical studies of efficacy and safety should be carried out using most of the other above-mentioned oral agents and these may also gain a place in the differential therapy of ED. There are no studies directly comparing sildenafil and other treatments of ED or assessing its role in combination with other therapies [10]. In patients without contraindications for its use, sildenafil is a highly effective treatment for ED, which can be considered as the first-line therapy in ED with an acceptable risk–benefit ratio. The impact of ED treatment with sildenafil on aspects of partnership and life satisfaction will be outlined in the following Chapters.

Psychosocial implications of ED

ED has multi-faceted psychological implications with regard to self-esteem, mood, and anxiety. In patients with ED, a vicious circle frequently develops: initially impaired erectile function is followed by lowered self-confidence in sexual performance and reduced self-esteem [11]. This leads to a heightened anxiety level, particularly high expectancy anxiety, and anticipatory ruminations and negative beliefs concerning sexual performance. Heightened tension and anxiety contribute to a further lowering of erectile performance and, as a

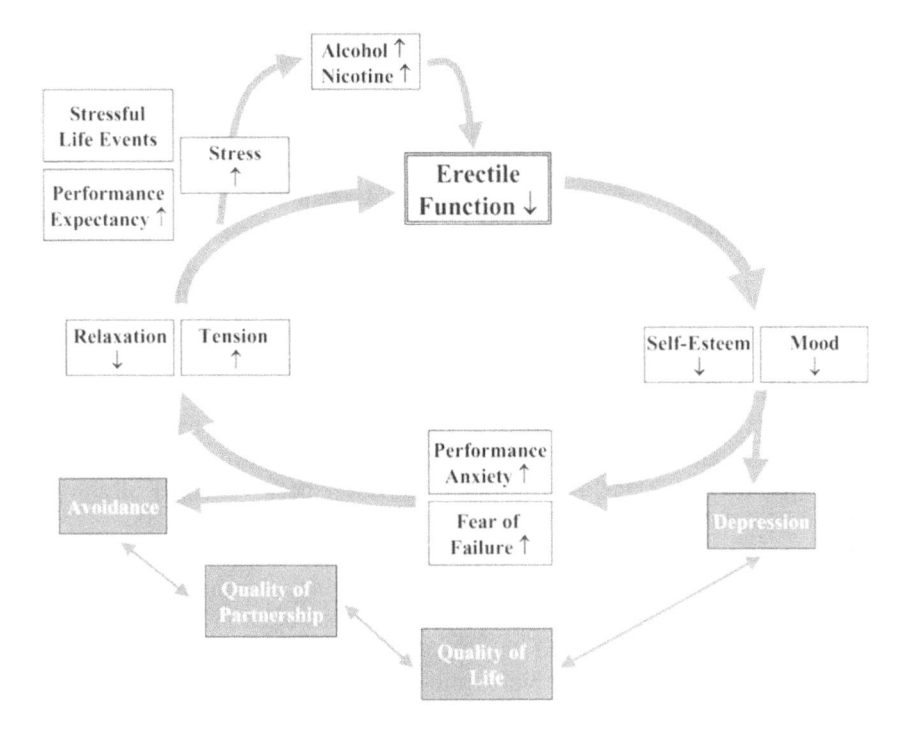

Figure 2. Erectile dysfunction and psychological implications.

consequence, sometimes lead to reduced sexual arousal. Over time, guilt feelings and avoidance behavior with respect to sexual activities can develop (Fig. 2).

Accordingly, the successful treatment of ED may not only improve sexual performance, but also prevent or counteract unfavorable psychosocial consequences and improve or restore quality of partnership (QoP) and quality of life (QoL). Successful treatment of ED comprises not only efficacy but also acceptance of the treatment by patients and their partners.

As already mentioned, sildenafil was the first effective oral treatment of ED with an essential impact on the management of ED [10]. However, before sildenafil was introduced as the first PDE5 inhibitor, there were also doubts and concerns assuming negative consequences of sildenafil treatment. It was surmised that female partners of successfully treated patients with ED could loose their feeling of attractiveness, because they might attribute the partner's sexual arousal to the effects of sildenafil. Another issue was put forth by psychoanalysts who argued that many partnerships are balanced and somehow stabilized by erectile problems. Effective pharmacological treatment of ED without psychotherapeutic support could, hence, lead to disturbances and confusion in the partnership. Therefore, the question of an impact of sildenafil treatment of ED on aspects of QoP and QoL has to be answered empirically.

Assessment methods for erectile function, quality of partnership, and quality of life

Following the model outlined in Figure 1, different methodological techniques and assessment instruments are used to assess erectile function, QoP, and QoL. For the assessment of erectile function and related sexual problems, the International Index of Erectile Function (IIEF) has been developed and validated parallel to the studies with sildenafil [12, 13]. The IIEF is currently the gold standard for assessment of erectile dysfunction and treatment-related changes.

Quality of partnership and quality of life are self-explanatory concepts with high face validity. However, when such apparently complex issues are a focus of empirical research, a theoretical framework and established assessment instruments could become necessary. In the domain of QoP, neither the concept has been elaborated nor has a detailed assessment instrument been established yet. Moreover, the aspect of QoP has rarely been systematically investigated including the female partners of ED patients.

One study with particular focus on that issue has used a partnership questionnaire (PFB), a differentiated instrument covering relevant aspects of partnership which was validated for both men and women [14–17].

Regarding QoL, general and illness-specific aspects are somehow arbitrarily distinguished (generic QoL and health-related QoL). The most widely used scale to assess both generic and illness-related aspects of QoL is the SF-36 scale [18] or its abbreviated form, the SF-12 [19]. One major advantage of the SF-36 is the large body of experience and a growing data base which allows comparisons across disorders and treatments. Other frequently used QoL scales are the Psychological General Well Being Index (PGWBI) [20], the Nottingham Health Profile (NHP), and the Sickness Impact Profile (SIP) [21].

However, there is a great variety of instruments used to assess QoL in studies on ED and sildenafil. So far, no standard for reporting on QoL aspects has been established [22]. Table 1 summarizes the most important instruments for the three domains of erectile function, QoP, and QoL, used in studies with sildenafil.

Overview of study results

Results from several studies investigating ED patients with various etiological factors have provided broad evidence regarding the improvement of erectile function, QoP, and QoL under treatment with sildenafil (Tab. 2). Typical and exemplary studies and results will be outlined in the following.

ED in patients with mixed etiology

When comparing patients with different etiologies of ED, it seems important to note, that despite differences in the response rates to sildenafil, there was not

Table 1. Self-administered assessment instruments for erectile function, QoP, and QoL

Instrument		Brief description and subscales	Domains	Ref.
International Index of Erectile Function	IIEF	Self administered, 15 questions Assessment of physiological and psychological aspects of erectile function in five domains: erectile function, orgasmic function, sexual desire, intercourse satisfaction, and overall satisfaction Standard assessment instrument for ED	EF (QoP) (QoL)	[12, 13]
Quality of Partnership – Partnerschaftsfragebogen	PFB	Self-administered questionnaire separate forms for male and female partners 30 items, 3 scales (10 items each): *Quarreling:* aggressive or quarreling behavior *Tenderness:* tenderness and intimacy *Togetherness:* togetherness and communication German instrument, validated in several studies on partnership problems and therapy	QoP	[14–15]
Short Form 36 for the assessment of Quality of Life	SF-36	Self-administered questionnaire several domains of generic QoL The SF-36 includes one multi-item scale measuring each of eight health concepts: 1. *Physical functioning* 2. *Role limitations due to physical health problems* 3. *Bodily pain* 4. *Social functioning* 5. *General mental health* 6. *Role limitations because of emotional problems* 7. *Vitality (energy/fatigue); and* 8. *General health perceptions* The SF-36 also includes a single-item measure of health transition or change. The SF-36 can also be divided into two aggregate summary measures, the Physical Component Summary (PCS) and the Mental Component Summary (MCS).	QoL	[18]
	SF-12	The SF-12 is a validated short form with 12 items SF-36 and SF-12 are QoL standard assessment instruments		[19]

Note: EF = erectile function; QoP = quality of partnership; QoL = quality of life

any single patient characteristic that could predict absolute failure for sildenafil therapy. One of the first follow-up studies on the subjective response to

Table 2. Improvement of QoL and QoP after treatment of ED with sildenafil

Study cohort	Design	Assessments	Reference
ED of different etiologies	open	IIEF, QoP, well-being	[43]
	open	IIEF, QoL, QoP, Depression	[16, 17, 20, 39]
	open	IIEF, QoL	[24]
	open	IIEF, QoP, self-esteem	[25]
	open	IIEF, QoL	[44]
	db, pl	IIEF, QoL, QoP	[45]
	db, pl	IIEF, QoP	[26]
ED + Ageing (Age > 65 years)	open	IIEF, QoL	[28]
ED + Diabetes mellitus	db, pl	IIEF, QoL	[29]
ED + Spinal cord injury	db, pl	IIEF, QoL, QoP	[30]
ED + Uremia, chronic dialysis	open	IIEF, QoP	[31]
	open	IIEF, QoL	[32]
ED in kidney transplant recipients	open	IIEF, QOL, QoP	[33]
ED + Parkinson's disease	db, pl	IIEF, QoL	[35]
ED + Mild-to-moderate Depression	db, pl	IIEF, QoL	[38]
	open	IIEF, QoL, QoP	[46]
ED + Prostate cancer therapy	Review	(IIEF, QoL)	[40]
	db, pl	IIEF (QoL)	[41]
	open	IIEF (QoL)	[42]

Note: ED = erectile dysfunction; IIED = international index of erectile dysfunction; QoL = quality of life; QoP = quality of partnership; db = double-blind; pl = placebo-controlled

sildenafil treatment in 267 patients with a mean age of 61 years and a duration of ED of at least four years showed that after six weeks treatment satisfaction was reported by 65% of the patients [23]. Even patients with severe ED and a neurogenic cause had a 41% satisfaction rate. Predictors of a more favorable response to sildenafil were thus non-neurogenic cause of ED and a better erectile function at baseline. However, improvement of ED was accompanied in all subgroups by improved QoL [23].

Additionally, recent studies have replicated the earlier findings that sildenafil treatment is well-tolerated. Only few patients discontinue treatment because of adverse events which is a prerequisite for an unbiased interpretation of study results on QoL measurements. The results from the sildenafil studies suggest that a particularly flexible dosing regimen with oral sildenafil starting with 25 mg is safe and has beneficial effects on all indices of erectile function and QoL [24].

More specifically, in a 10-week open-label study with 93 ED patients it was recently shown that improved erections were in fact paralleled by improvement of sexual relationship, confidence, and self-esteem [25]. Correlations between an improvement of IIEF scores and positive changes in self-esteem and QoP were in the range of 0.45–0.65. Accordingly, a substantial proportion of the variance in QoL and QoP was explained by improved erectile function. However, other treatment-related factors that might indirectly influence QoL and QoP are nevertheless important (c.f. Figs 1 and 2). A large-scale, double-blind, placebo controlled study with 247 ED patients and their female partners, confirmed and extended these findings: sildenafil was highly effective in improving erectile function after 12-weeks flexible dosing, and this effect was corroborated by partner evaluations using a standardized assessment scale [26].

It can be assumed that – similarly to the development of a vicious circle – an advantageous feedback process can be started by successful treatment of ED leading to relaxed sexual behavior, regained self-esteem, and improved partnership. The findings of improved QoP in patients treated with sildenafil (n = 53) compared to a group of patients and partners awaiting sildenafil treatment (n = 51) corroborates these assumptions [16, 17]. Patients and their partners were separately investigated using the PFB for QoP. The groups (sildenafil *versus* no treatment) were comparable in demographic and illness variables, but differed significantly with respect to QoP (PFB) as perceived by patients and their female partners. In men, all three PFB subscales (quarreling, tenderness, togetherness) differed significantly between the two study groups. In the female partners the domains tenderness and togetherness differed significantly. Figure 3 shows the results for the tenderness subscale. Erectile function was correlated highly significantly with tenderness and togetherness in patients and their partners, i.e., improved erectile function was accompanied by a heightened feeling of tenderness not only in the treated patients, but similarly also in their female partners.

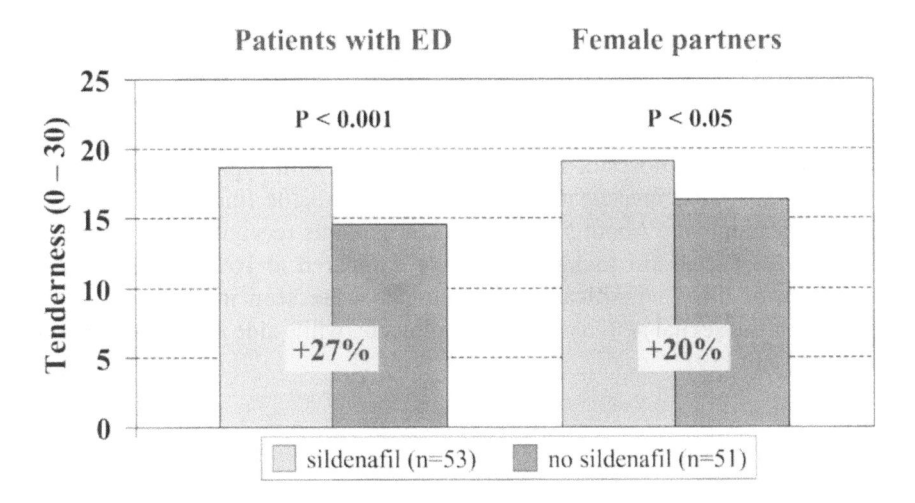

Figure 3. Quality of partnership (tenderness) in ED patients with or before treatment with sildenafil and their female partners. Note: The results are from [16, 17]

Ageing male

Age-related changes in men have recently received widespread attention. Sexual and partnership satisfaction are reduced in men who were dissatisfied with their health. Somatic and psychosocial factors contribute to findings of an age-related decline in sexual satisfaction, and a slight increase in partnership satisfaction [2]. Despite the knowledge from epidemiological studies (MMAS) [27] that ED is age-related possibly involving low testosterone levels and other factors, only a few ED studies have focused on ageing males. In an open study with 150 patients below 65 years and 44 patients over 65 years, sildenafil was used to treat ED. Sildenafil efficacy in the younger group (89.1%) was significantly greater than in the older group (65.7%) whereas serum testosterone and prolactin levels were similar between groups. Older patients, however, showed a higher prevalence of diabetes mellitus, hypertension, and benign prostate hyperplasia.

Only diabetes appeared to decrease the efficacy of sildenafil significantly in older patients, although efficacy in older patients even without diabetes was relatively low. Nevertheless, nearly two-thirds of the older men had a good response to the drug and reported improved QoL. Additionally, no adverse drug effect was specific to older ED patients [28].

The promising results of sildenafil treatment in ED patients with psychogenic and mixed etiology were extended by studies in patient groups suffering from ED due to a specific illness or medical condition.

Diabetes mellitus

In patients with diabetes mellitus, ED is a long-known and common complication emerging in 30–80% of patients. Several studies have shown that sildenafil is safe and effective in patients with diabetes mellitus, although response rates are moderately lower in some studies compared to ED patients without diabetes. In a multicenter, randomized, double-blind, placebo-controlled study with flexible doses of sildenafil, 92 patients with Type II diabetes mellitus and ED were investigated with respect to erectile function (IIEF) and QoL measurements. About 55% of diabetic patients receiving sildenafil had at least one successful sexual intercourse compared to 16% in the placebo group. In addition, significant improvements were seen in QoL and other aspects of the sexual activity of treated subjects, while side effects were mild and transient [29].

Spinal cord injury

ED in patients with spinal cord injuries has often been regarded as treatment-resistant in the past. Since the advent of sildenafil for oral treatment, several case reports have shown that even in this patient group successful treatment of ED is possible.

In a multicenter, randomized, double-blind, placebo-controlled, flexible-dose, crossover study the effect of sildenafil on QoL of 178 men with ED caused by spinal cord injury was evaluated. QoL aspects were assessed with the SF-12, the PGWBI, and additional instruments. Significant improvements were seen for overall satisfaction with sex life, sexual relationship with partner, and concerns about erectile problems. Significant improvements (sildenafil *versus* placebo) were also reported for QoL parameters (mental health, wellbeing, depression, and anxiety) [30].

Uremia, chronic dialysis, and kidney transplant recipients

According to the literature, ED is present in approximately 30% of male patients with chronic renal failure and in 50% of patients undergoing hemodialytic treatment [31]. Reviews report a prevalence rate for ED in dialysis and transplant patient in the range between 17–82% [32]. Despite hemodialysis, fertility, libido and erectile dysfunction are often progressively impaired with time. ED is one of the main factors negatively influencing the QoL in patients with hemodialysis. Sildenafil treatment was effective in several studies not only with respect to erectile function, but also regarding subjective wellbeing and QoL. In one study with 20 patients aged 29–51 years and hemodialytic treatment history of 3–13 years, after three months of flexible treatment with sildenafil 25–50 mg all patients reported improved sexual activity and sexual desire as well as improved aspects of QoL [31]. Another study on 50 men after kidney transplantation replicated these findings [33].

Sildenafil in flexible dosing (25–100 mg) not only improved the patients' erection ability and the frequency of their erection maintenance significantly after six months. Additionally, 66% of patients reported that treatment had improved their erections, their sexual life, and their partner relationships. Notably, there were no interactions between sildenafil and the immunosuppressive drugs and there was no significant adverse effect of sildenafil on graft function [33].

Parkinson's disease

According to clinical observations, ED is highly prevalent in patients with Parkinson's disease (PD) and multiple system atrophy (MSA) [34]. However, controlled studies on QoL and QoP are rare. In one randomized, double-blind, placebo-controlled, crossover study, the efficacy and safety of sildenafil in 24 men with erectile dysfunction and Parkinsonism due either to PD or MSA was evaluated using the IIEF and QoL questionnaires [35]. The results demonstrated a clear effect of sildenafil 50 mg on the ability to achieve and maintain an erection and improvement in QoL. In three MSA patients, a severe drop in blood pressure one hour after taking sildenafil was reported; however, all patients with MSA reported good erectile response and were reluctant to discontinue the medication [35].

Mild-to-moderate depression

According to the Massachusetts Male Ageing Study (MMAS) [27] there is an age-corrected 1.8-fold risk for depressed men to develop ED. On the other hand, in patients with ED, depressive symptoms are common (c.f. Fig. 2) [36, 37]. Adequate antidepressant therapy can sometimes improve ED, but can also lead to newly emerging sexual dysfunction.

Satisfactory treatment of ED, however, has been shown to reduce depressive symptoms in clinically depressed patients [38]. In a well-conducted 12-week, randomized, double-blind, placebo-controlled trial 152 men with ED (mean duration of ED 5.7 years; mean age 56 years) and a psychiatric diagnosis of depressive disorder were assigned to flexible-dose treatment with sildenafil or placebo. Observer-rated and self-report instruments were used to assess changes in sexual function, depressive symptoms, and QoL. The study impressively showed that sildenafil treatment significantly improved erectile function (responders: 73% sildenafil *versus* 14% placebo). 76% of treatment responders showed an at least 50% improvement in Hamilton depression scale scores *versus* 14% of non-responders. QoL was also substantially improved in treatment responders and correlated with improved erectile function (IIEF). The authors concluded that sildenafil is efficacious for ED in men with mild-to-moderate depressive illness and that improvement of ED is associated with marked improvement in depressive symptoms and QoL [38].

Moreover, sildenafil treatment improved wellbeing, positive mood aspects, and QoL in non-depressed men with ED [39]. In this study, depression scores were analyzed in 53 patients after at least four weeks of sildenafil treatment, and in 51 patients with ED awaiting sildenafil treatment (mean age 56 years). The open-label, cross-sectional study showed a significant improvement of positive mood aspects in patients treated with sildenafil. The study underlines the relevance of depression and the importance of effective ED treatment. Although depression was generally low in this sample, hedonistic aspects of wellbeing and QoL were substantially enhanced in the group of ED patients after sildenafil treatment (Fig. 4).

Prostate cancer therapy

According to recent studies, erectile dysfunction following prostate brachytherapy for prostate cancer is common with approximately 50% of patients developing ED within five years of implantation. Several factors

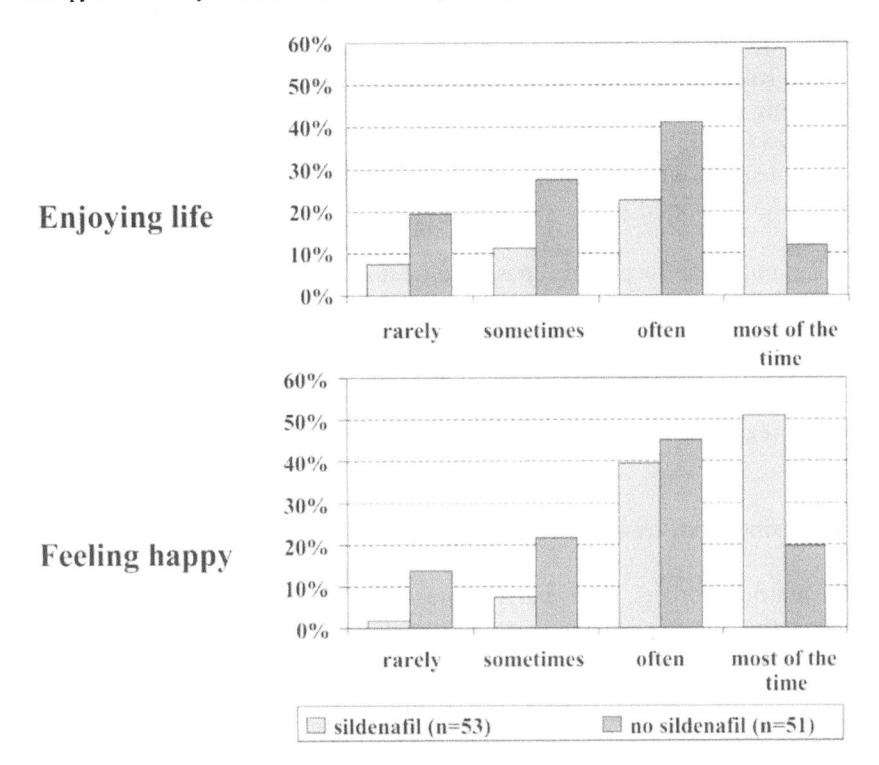

Figure 4. Enhanced positive mood aspects after sildenafil treatment of ED. Note: Differences between groups were highly statistically significant (Chi^2-tests; $p < 0.0001$). Data from [39].

including pre-implant potency, patient age, radiation, and diabetes mellitus additionally exaggerate brachytherapy-related ED. The majority of patients with brachytherapy-induced ED showed a response to sildenafil in terms of erectile function and improved quality of life [40]. Similar findings come from a double-blind study [41] with a two-year open-label extension in 60 patients treated with radiotherapy for prostate cancer followed by ED [42].

In most of the patients sildenafil caused a significant improvement in IIEF scores. About 25% of the patients still used sildenafil two years after trial entry [42] which is at least an indirect evidence for a subjective improvement also.

Table 2 (see p. 88) summarizes the results of studies with assessment of QoL and QoP in patients with sildenafil treatment for ED.

Radical prostatectomy and treatment of rectal cancer and inflammatory bowel disease

Due to pelvic parasympathetic nerve damage, ED in patients with rectal excision for rectal cancer and inflammatory bowel disease is not rare. Treatment with sildenafil was shown to be effective in a randomized, double-blind, placebo-controlled study with 79 men after proctectomy [47]. The prevalence of ED following radical prostatectomy is very high and a well recognized source of impaired QoL in these patients [48]. However, research in this field including assessment of QoL and QoP is still rare, and patients complain about a lack of professional support [48].

The mutual influence of illness and treatment on subjective wellbeing and QoL is well known (see above). However, an increasingly important clinical problem is the negative impact of highly effective pharmacological treatments on sexual function. Beside medications for endocrine, cardiovascular and neurological diseases, psychopharmacological treatments are at highest risk to induce sexual disturbances on different levels.

ED arising from treatment with psychotropic drugs

Psychiatric disorders, particularly depressive disorders and schizophrenia, are frequently connected with sexual dysfunction including ED. Unfortunately, treatment with most antidepressants and neuroleptics is also associated with impaired sexual functioning [49, 50]. There is no doubt, that sexual side effects may reduce quality of life and may give rise to non-compliance [51]. Management of sexual side effects induced by psychotropic drugs comprises a reduction in dosage, switching to another medication, combination with different psychotropic agents, or specific treatment of sexual dysfunction such as sildenafil in case of ED.

Despite the lack of controlled studies, there is emerging evidence that sildenafil and other PDE5 inhibitors can be very helpful in treatment of ED arising under treatment with psychotropic medication.

Negative effects of sildenafil treatment

Despite several warnings with respect to moral, familiar, and partnership aspects before sildenafil was marketed, only a few such non-pharmacological "side effects" have been reported after the introduction and widespread use of the PDE5-inhibitors, if contraindications and necessary precautions are recognized. Two cases of 70 years old men who have been transferred to a gerontopsychiatric ward after intake of sildenafil have been reported in 2000 [52]. In one case the patient has developed aggressive outbursts after sildenafil intake in response to rejection by his wife following increased sexual desire of the treated patient. The other patient (ED and diabetes mellitus) developed delusions of jealousy after unsuccessful treatment of the ED with sildenafil.

Summary and perspectives

Erectile dysfunction (ED) is a highly prevalent and still undertreated condition. It may be a symptom of underlying, chronic illness and can contribute to anxiety, depression and loss of self-esteem with a negative impact on quality of life, psychosocial health, and relationships [53, 54]. The aging of the population, as well as the introduction of new treatment options, such as sildenafil, has led to increased public awareness of this disorder. New oral therapeutic agents are marketed or on the horizon [4].

The results of various studies have confirmed the efficacy of sildenafil in men with ED of different etiologies, as well as the positive effect of sildenafil on the quality of partnership and quality of life (Fig. 5).

According to the present knowledge, the quality of life, not only of patients but also of their sexual partners, is in most cases improved significantly with sildenafil usage and this is an important precondition for overall health of both.

However, QoP is still a widely underreported aspect of QoL. The majority of studies do not include a specific assessment of QoP. Published data indicate lower levels of intimacy in partnerships with sexually dysfunctional men [55]. Carroll has reported [56] that in most sexual relationships men are the initiators and that their withdrawal due to ED often leads to reduced sexual activities of the couple. According to this study, 41% of women in such partnerships are feeling responsible and to blame for the sexual difficulties.

Concerns prior to the release of sildenafil to the market – that an erection induced by therapeutic intervention may lead to resistance in the female partners – were not confirmed by empirical data [57]. In all studies including

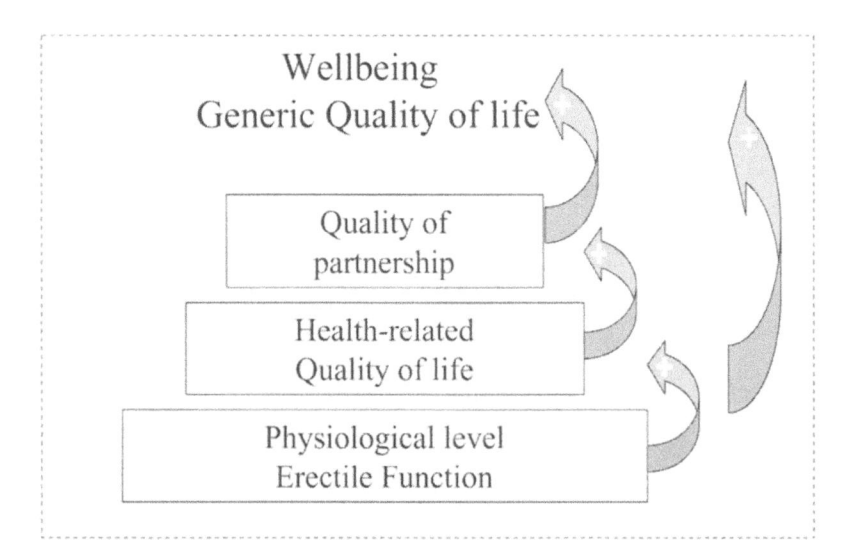

Figure 5. Positive impact of improved ED on quality of partnership, quality of life, and wellbeing.

patients with difficult-to-treat medical conditions with assessment of QoP, sildenafil not only enhanced erectile function, but also improved partnership.

Moreover, improvement of ED as measured with the IIEF correlated strongly with the improvement in partnership parameters, i.e., the better the outcome regarding ED, the better the QoP was. Exceptions with negative therapeutic effects were reported only very rarely in patients with (mis-)use of sildenafil [52].

Interestingly, the effects of sildenafil treatment on QoP are in line with earlier studies using other treatment options for ED, e.g., vacuum erection devices [58, 59], self-injection therapy [60], and transurethral alprostadil [61]. In summary, all studies so far report beneficial effects of successful ED treatment on partnership. However, only few studies comprehensively addressed the various aspects of dyadic interaction and its relation to ED treatment. Additionally, studies with combined pharmacological and psychotherapeutical treatment are grossly lacking. It is increasingly recognized that ED usually arises from mixed organic and psychogenic causes, but the management of this condition often neglects the complexity of most cases of ED.

While therapy with sildenafil and other PDE5 inhibitors can play an important role in many cases of ED, physicians should recognize and try to address the psychological and interpersonal context of ED in their patients [62].

The results of sildenafil treatment of ED on QoL aspects are also impressive. In most studies with a broad variety of patients' characteristics, sildenafil had a substantial positive impact on QoL. Placebo-controlled studies showed that improvement of QoL under sildenafil offers an increase in QoL parameters of 20–80% compared to placebo. However, these results should be interpreted appropriately. There is still no globally accepted standard of QoL measurement [21, 22]. Almost all studies used the IIEF, which was developed to assess ED and changes in erectile function [12, 13]. However, beside so-called health-related or illness-specific QoL assessments (e.g., [20]), a diversity of generic or global QoL measurements are used. The most frequently used QoL scale is the SF-36 or its abbreviated version, the SF-12 [18, 19]. Comparisons of effects derived from the SF-36 scale are hardly comparable to other scales making conclusions difficult.

Public acceptance and cost-effectiveness of sildenafil treatment and quality of life

Despite all the evidence for positive effects of successful ED treatment with sildenafil on QoP and QoL, the question in the public opinion remains whether or not such a treatment is cost-effective and necessary.

An empirical study [63] asked the interesting question if patients and the general public are like-minded about the effect of ED on quality of life. The background of this question was the assumption that patients with ED – in order to emphasize claims for treatment – may overestimate treatment effects. In funding debates the question becomes important whether the general public agrees on the value of a treatment. Consequently, the authors studied the social values for ED treatments in 106 ED patients and 169 representative control subjects. The result was surprising as both patients and the controls from general population considered ED similarly an important aspect of QoL [63]. The authors concluded that funding for ED treatment might be considered. Moreover, because the value for erectile function was equivalent in different age groups, there might be no convincing argument to limit funding to young patients.

Cost-effectiveness [64] of sildenafil treatment for ED was investigated in some studies comparing sildenafil to other therapies or no-drug-therapy with "costs per quality-adjusted life-year (QALY) gained" serving as outcome parameter [65, 66]. Sildenafil treatment compared favorably with that of accepted therapies for other medical conditions [65]. For men whose QoL is substantially diminished by impotence, sildenafil was considered cost-effective relative to other commonly used health interventions [66].

In a cost-utility analysis of sildenafil compared with papaverine–phentolamine injections, costs and effects of a clinical trial were converted into "quality adjusted life years" (QALYs) [67]. Overall, treatment with sildenafil gained more QALYs, but the total costs were higher. In summary, treatment with sildenafil is cost-effective; however, when considering funding sildenafil, healthcare systems should take into account that the frequency of use strongly affects cost-effectiveness [67].

Final comments

To ensure positive effects on ED, QoP and QoL when taking sildenafil, some precautions and contraindications have to be taken into account. Given the incidence of unknown cardiovascular disease in patients with ED and the indications and contraindications of PDE5 inhibitors in patients with cardiovascular diseases, use and contraindications should be considered in all patients with ED [68, 69].

The responsibilities of treating patients with ED include educating the patient about sexually transmitted diseases, providing general sex education and counseling to the patient and his partner, and providing a treatment that is appropriate for the cause of the problem [70].

Sildenafil has changed private and social life in patients with ED and their partners all over the world. However, whether or not this treatment is used remains the question of individual choice [71].

Acknowledgements
Data in Figure 3 and Figure 4 are taken from a study [16, 17, 20, 39] supported by Pfizer Inc.

References

1 Maslow AH (1954) *Motivation and personality*. New York, Harper
2 Beutel ME, Schumacher J, Weidner W, Brähler E (2002) Sexual activity, sexual and partnership satisfaction in ageing men – results from a German representative community study. *Andrologia* 34: 22–28
3 Walton B, Thorton T (2003) Female sexual dysfunction. *Curr Womens Health Rep* 3: 319–326
4 Shabsigh R, Anastasiadis AG (2003) Erectile dysfunction. *Annu Rev Med* 54: 153–168
5 Dunn KM, Croft PR, Hackett GI (1999) Association of sexual problems with social, psychological, and physical problems in men and women: a cross sectional population survey. *J Epidemiol Community Health* 53: 144–148
6 Schilling G, Müller MJ, Haidl G (1999) Sexual dissatisfaction and somatic complaints in male infertility. *Psychother Psychosom Med Psychol* 49: 256–263
7 Vitezic D, Pelcic JM (2002) Erectile dysfunction: oral pharmacotherapy options. *Int J Clin Pharmacol Ther* 40: 393–403
8 Langtry HD, Markham A (1999) Sildenafil: a review of its use in erectile dysfunction *Drugs* 57: 967–989
9 McMahon CG, Samali R, Johnson H (2000) Efficacy, safety and patient acceptance of sildenafil citrate as treatment for erectile dysfunction. *J Urol* 164: 1192–1196
10 Salonia A, Rigatti P, Montorsi F (2003) Sildenafil in erectile dysfunction: a critical review. *Curr Med Res Opin* 19: 241–262
11 Hartmann U (1998) Psychological stress factors in erectile dysfunctions. Causal models and empirical results. *Urologe A* 37: 487–494
12 Rosen RC, Riley A, Wagner G, Osterloh IH, Kirkpatrick J, Mishra A (1997) The international index of erectile function (IIEF): a multidimensional scale for assessment of erectile dysfunction. *Urology* 49: 822–830
13 Rosen RC, Cappelleri JC, Gendrano N 3rd (2002) The international index of erectile function (IIEF): a state-of-the-science review. *Int J Impot Res* 14: 226–244
14 Hahlweg K (1988) Partnership questionnaire PFB. In: M Hersen, AS Bellack (eds): *Dictionary of behavioral assessment devices*. Pergamon, New York, 337–339
15 Hahlweg K (1996) *Fragebogen zur Partnerschaftsdiagnostik*. Handanweisung. Hogrefe Verlag GmbH, Göttingen
16 Müller MJ (2001) Die Beziehung profitiert signifikant. Eine ED-Pilotstudie untersuchte, ob sich Lebens- und Partnerschaftsqualität unter Sildenafil-Therapie verbessern. *Urologische Nachrichten* 08: 10
17 Müller MJ, Ruof J, Graf-Morgenstern M, Porst H, Benkert O (2001) Quality of partnership in erectile dysfunction patients responsive to oral sildenafil treatment *vs* untreated patients: results of a pilot study. *Pharmacopsychiatry* 34: 91–95
18 Ware JE Jr, Kosinski M, Keller SD (1994) *SF 36 Physical and Mental Health Summary Scales: A User's Manual*. MA: The Health Institute, New England Medical Center, Boston
19 Ware JE Jr, Kosinski M, Keller SD (1996) A 12 Item Short Form Health Survey: Construction of scales and preliminary tests of reliability and validity. *Med Care* 34: 220–233
20 Ruof J, Graf-Morgenstern M, Müller MJ (2001) Lebensqualität bei Patienten mit erektiler Dysfunktion (ED). *Aktuelle Urologie* 32: 21–26
21 Gill TM, Feinstein AR (1994) A critical appraisal of the quality of quality-of-life measurements. *JAMA* 272: 619–626
22 Sanders C, Egger M, Donovan J, Tallon D, Frankel S (1998) Reporting on quality of life in randomised controlled trials: biblographic study. *BMJ* 317: 1191–1194
23 Jarow JP, Burnett AL, Geringer AM (1999) Clinical efficacy of sildenafil citrate based on etiology and response to prior treatment. *J Urol* 162: 722–725
24 Benchekroun A, Faik M, Benjelloun S, Bennani S, El Mrini M, Smires A (2003) A baseline-controlled, open-label, flexible dose-escalation study to assess the safety and efficacy of sildenafil citrate (Viagra) in patients with erectile dysfunction. *Int J Impot Res* 15 Suppl 1: S19–24
25 Althof SE, Cappelleri JC, Shpilsky A, Stecher V, Diuguid C, Sweeney M, Duttagupta S (2003) Treatment responsiveness of the Self-Esteem and relationship questionnaire in erectile dysfunction. *Urology* 61: 888–892
26 Lewis R, Bennett CJ, Borkon WD, Boykin WH, Althof S, Stecher VJ, Siegel RL (2001) Patient and

partner satisfaction with Viagra (sildenafil citrate) treatment as determined by the Erectile Dysfunction Inventory of Treatment Satisfaction Questionnaire. *Urology* 57: 960–965
27 Araujo AB, Durante R, Feldman HA, Goldstein I, McKinlay JB (1998) The relationship between depressive symptoms and male erectile dysfunction: cross-sectional results from the Massachusetts Male Aging Study. *Psychosom Med* 60: 458–465
28 Tsujimura A, Yamanaka M, Takahashi T, Miura H, Nishimura K, Koga M, Iwasa A, Takeyama M, Matsumiya K, Takahara S et al. (2002) The clinical studies of sildenafil for the ageing male. *Int J Androl* 25: 28–33
29 Escobar-Jimenez F, Grupo de Estuido Espanol sobre Sildenafilo (2002) Efficacy and safety of sildenafil in men with type 2 diabetes mellitus and erectile dysfunction. *Med Clin* 119: 121–124
30 Hultling C, Giuliano F, Quirk F, Pena B, Mishra A, Smith MD (2000) Quality of life in patients with spinal cord injury receiving Viagra (sildenafil citrate) for the treatment of erectile dysfunction. *Spinal Cord* 38: 363–370
31 Chen J, Mabjeesh NJ, Greenstein A, Nadu A, Matzkin H (2001) Clinical efficacy of sildenafil in patients on chronic dialysis. *J Urol* 165: 819–821
32 Yenicerioglu Y, Kefi A, Aslan G, Cavdar C, Esen AA, Camsari T, Celebi I (2002) Efficacy and safety of sildenafil for treating erectile dysfunction in patients on dialysis. *BJU Int* 90: 442–445
33 Barrou B, Cuzin B, Malavaud B, Petit J, Pariente JL, Buchler M, Cormier L, Benoit G, Costa P (2003) Early experience with sildenafil for the treatment of erectile dysfunction in renal transplant recipients. *Nephrol Dial Transplant* 18: 411–417
34 Lüders M, Boxdorfer S, Beier KM (1999) Partnership and sexuality in cases of Parkinson's disease. *Sexuologie* 6: 18–29
35 Hussain IF, Brady CM, Swinn MJ, Mathias CJ, Fowler CJ (2001) Treatment of erectile dysfunction with sildenafil citrate (Viagra) in parkinsonism due to Parkinson's disease or multiple system atrophy with observations on orthostatic hypotension. *J Neurol Neurosurg Psychiatry* 71: 371–374
36 Holden RR (1999) The Holden psychological screening inventory and sexual efficacy in urological patients with erectile dysfunction. *Psychol Rep* 84: 255–258
37 Shabsigh R, Klein LT, Seidman S, Kaplan SA, Lehrhoff BJ, Ritter JS (1998) Increased incidence of depressive symptoms in men with erectile dysfunction. *Urology* 52: 848–852
38 Seidman SN, Roose SP, Menza MA, Shabsigh R, Rosen RC (2001) Treatment of erectile dysfunction in men with depressive symptoms: results of a placebo-controlled trial with sildenafil citrate. *Am J Psychiatry* 158: 1623–1630
39 Müller MJ, Benkert O (2001) Lower self-reported depression in patients with erectile dysfunction after treatment with sildenafil. *J Affect Dis* 66: 255–261
40 Merrick GS, Wallner KE, Butler WM (2003) Management of sexual dysfunction after prostate brachytherapy. *Oncology* 17: 52–62
41 Incrocci L, Koper PC, Hop WC, Slob AK (2003) Sildenafil citrate and erectile dysfunction following external beam radiotherapy for prostate cancer: a randomized, double-blind, placebo-controlled, cross-over study. *Urology* 62: 116–120
42 Incrocci L, Hop WC, Slob AK (2003b) Efficacy of sildenafil in an open-label study as a continuation of a double-blind study in the treatment of erectile dysfunction after radiotherapy for prostate cancer. *Urology* 62: 116–120
43 Paige NM, Hays RD, Litwin MS, Rajfer J, Shapiro MF (2001) Improvement in emotional well-being and relationships of users of sildenafil. *J Urol* 166: 1774–1778
44 Fujisawa M, Sawada K, Okada H, Arakawa S, Saito S, Kamidono S (2002) Evaluation of health-related quality of life in patients treated for erectile dysfunction with Viagra (sildenafil citrate) using SF-36 score. *Arch Androl* 48: 15–21
45 Giuliano F, Pena BM, Mishra A, Smith MD (2001) Efficacy results and quality-of-life measures in men receiving sildenafil citrate for the treatment of erectile dysfunction. *Qual Life Res* 10: 359–369
46 Martinez-Sanchez E, Oyaguez Martin I, Carrasco Garrido P, Gil de Miguel A (2001) Treatment with sildenafil and satisfaction with life in depressive patients with erectile dysfunction. *Actas Esp Psyquiatr* 29: 293–298
47 Lindsey I, George B, Kettlewell M, Mortensen N (2002) Randomized, double-blind, placebo-controlled trial of sildenafil for erectile dysfunction after rectal excision for cancer and inflammatory bowel disease. *Dis Colon Rectum* 45: 727–732
48 Meuleman EJ, Mulders PF (2003) Erectile function after prostatectomy: a review. *European*

Urology 43: 95–101
49 Kristensen E (2002) Sexual side effects induced by psychotropic drugs. *Dan Med Bull* 49: 349–352
50 Zajecka J (2001) Strategies for the treatment of antidepressant-related sexual dysfunction. *J Clin Psychiatry* 62 Suppl 3: 35–43
51 Wallace M (2001) Real progress-the patient's perspective. *Int Clin Psychopharmacol* 16 Suppl 1: S21–S24
52 Mintz D (2000) Unusual case report: Nonpharmacologic effects of sildenafil. *Psychiatr Serv* 51: 674–675
53 Levine LA (2000) Diagnosis and treatment of erectile dysfunction. *Am J Med* 109 Suppl 9A: 3S–12S
54 Ventegodt S (1998) Sex and quality of life in Denmark. *Arch Sex Behav* 27: 295–307
55 McCabe MP (1997) Intimacy and quality of life among sexually dysfunctional men and women. *J Sex Marital Ther* 23: 276–290
56 Carroll JL, Bagley DH (1990) Evaluation of sexual satisfaction in partners of men experiencing erectile failure. *J Sex Marital Ther* 16: 70–78
57 Hultling C (1999) Partners' perceptions of the efficacy of sildenafil citrate (VIAGRA) in the treatment of erectile dysfunction. *Int J Clin Pract Suppl* 102: 16–18
58 Turner LA, Althof SE, Levine SB, Bodner DR, Kursh ED, Resnick MI (1991) External vacuum devices in the treatment of erectile function: a one year study of sexual and psychosocial impact. *J Sex Marital Ther* 17: 81–93
59 Denil J, Ohl DA, Smythe C (1996) Vacuum erection device in spinal cord injured men: patient and partner satisfaction. *Arch Phys Med Rehabil* 77: 750–753
60 Althof SE, Turner LA, Levine SB, Bodner D, Kursh ED, Resnick MI (1992) Through the eyes of women: the sexual and psychological responses of women to their partner's treatment with self-injection or external vacuum therapy. *J Urol* 147: 1024–1027
61 Williams G, Abbou CC, Amar ET, Desvaux P, Flam TA, Lycklama GAB, Nijeholt A, Lynch SF, Morgan RJ, Müller SC et al. (1998) The effect of transurethral alprostadil on the quality of life of men with erectile dysfunction and their partners. *Br J Urol* 82: 847–854
62 Levine SB (2003) Erectile dysfunction: why drug therapy isn't always enough. *Cleve Clin J Med* 70: 241–246
63 Stolk EA, Busschbach JJ (2003) Are patients and the general public like-minded about the effect of erectile dysfunction on quality of life? *Urology* 61: 810–815
64 Spilker B (ed) (1996) *Quality of Life and Pharmacoeconomics in Clinical Trial. 2nd edition.* Lippincott-Raven Publishers, Philadelphia
65 Smith KJ, Roberts MS (2000) The cost-effectiveness of sildenafil. *Ann Intern Med* 132: 933–937
66 Kwok YS, Kim C (1999) Valuing Viagra: what is restoring potency worth? *Eff Clin Pract* 2: 171–175
67 Stolk EA, Busschbach JJ, Caffa M, Meuleman EJ, Rutten FF (2000) Cost utility analysis of sildenafil compared with papaverine-phentolamine injections. *BMJ* 320: 1165–1168
68 Görge G, Flüchter S, Kirstein M, Kunz T (2003) Sex, erectile dysfunction, and the heart: a growing problem. *Herz* 28: 284–290
69 Bedell SE, Graboys TB, Duperval M, Goldberg R (2002) Sildenafil in the cardiologist's office: patients' attitudes and physicians' practices toward discussions about sexual functioning. *Cardiology* 97: 79–82
70 Sadovsky R (2000) Integrating erectile dysfunction treatment into primary care practice. *Am J Med* 109 Suppl 9A: 22S–28S
71 Umrani DN, Goyal RK (1999) Pharmacology of sildenafil citrate. *Indian J Physiol Pharmacol* 43: 160–164

Erectile dysfunction, depression, and pharmacological treatments: biologic interactions

Stuart N. Seidman

Department of Psychiatry, College of Physicians and Surgeons of Columbia University; and the New York State Psychiatric Institute, 1051 Riverside Drive, Unit 98, New York, NY 10032, USA

Introduction

Sexual dysfunctions are common, in general, and especially prevalent among patients with psychiatric disorders. Male sexual dysfunction is associated with a range of imputed neurobiological, medical, psychological and interpersonal causes, and can have a significant impact on interpersonal functioning and quality of life. Moreover, the interaction between co-morbid sexual and psychiatric disorders has clinical implications for diagnosis, course, and treatment of both conditions. In recent years, there has been significant progress in the basic research and clinical therapeutics of male sexual dysfunction, with major advances in our understanding of the mechanisms of sexual response and the effects of pharmacologic agents on erectile function and orgasm. Here, we will briefly review the psychobiology of male sexual function and dysfunction, and describe diagnostic and therapeutic approaches; and then we will consider the interaction between erectile dysfunction and depression, with a focus on pharmacological treatments.

Normal sexual function

The sexual response cycle

There are four overlapping phases of sexual function: 1) drive; 2) arousal, marked by erection in men; 3) climax, marked by orgasm and ejaculation; and 4) resolution, which involves some degree of refractoriness. Normal sexual function is a biopsychosocial process. Sexual dysfunction commonly derives from the biological, psychological or social arena, but virtually always affects all three.

Age is an important determinant of sexual function. The change in sexual function with age is multi-factorial and variable [1]. Factors with particular (although not exclusive) relevance to age-related sexual problems include availability and health of a partner, relationship dynamics, fear of performance failure, chronic illness, substance abuse (particularly alcohol), medication side effects, medical morbidity (e.g., neuropathy, vascular insufficiency, hypogonadism/menopause), and depression [2].

The psychobiology of sexual function

Normal sexual behavior involves the complex interplay of endocrine, sensory, autonomic, cortical, and limbic neurophysiology. A comprehensive review of the psychobiology of sexual function is beyond the scope of this Chapter, and has been well-reviewed by Meston and Frohlich [3]. and Pfaus [4]. Here, we highlight central nervous system (CNS) modulators of sexual physiology, including monoamines and androgens, and the neurovascular physiology of erection.

Nervous system

Monoamines and sexual function
In humans, strong evidence for the role of monoamines in sexual function is derived from the side effect profiles of psychotropic medications, particularly antidepressants [5]. This is discussed in sections below, and here we highlight only the direction of effects supported by the literature. It should be recognized that most neurotransmitters with central effects on sexual function also influence the peripheral nervous system control of genital response. Moreover, there is a balance of interactions – central and peripheral – between neurotransmitters, such as monoamines, and other neuromodulators, such as nitric oxide (NO) and sex steroids. Localizing the sexual effects of neuropharmacologic manipulations is therefore difficult.

Catecholamines appear to have stimulatory effects on sexual behavior. It was reported in the 1970s that dopaminergic drugs given to Parkinsonian patients appeared to be pro-sexual, and clinical consensus is that dopamine agonists increase sexual desire and erectile function. Dopamine release in the medial preoptic area (MPOA) appears to stimulate increased copulatory rate and efficiency. Notably, the presence of testosterone has been shown to be permissive for dopamine release in the MPOA, probably mediated through NO production [6]. Dopamine antagonists inhibit sexual functioning, particularly orgasm. Norepinephrine (NE) levels increase during sexual activity, and are positively correlated with arousal – increasing dramatically at orgasm, and then decreasing precipitously (e.g., to baseline by 2 min post-orgasm).

In contrast to the activating effects of catecholamines, serotonin tends to inhibit sexual behavior. In monkeys, 5HT-1C/1D agonists inhibit sexual behavior. In male rats, activation of the 5HT-1A receptor lowers the threshold for ejaculation, while activation of 5HT-1B and 5HT-1C, and antagonism of 5HT-2, all inhibit sexual behavior [4].

Neurovascular physiology of erectile function

Basic research on neurovascular mechanisms has contributed greatly to our understanding of normal and pathological processes of erection. Smooth muscle contractility is regulated by a delicate balance between neurotransmitter and vasoactive substances mediating erectile tissue contraction (consistent with flaccidity) and relaxation (consistent with erection). NO plays a major role in inducing trabecular smooth muscle relaxation. Following release NO diffuses across smooth muscle cell membranes and binds to soluble guanylate cyclase. The resultant increase in intracellular cyclic guanosine monophosphate (cGMP) leads to binding of cGMP to cGMP-dependent protein kinases (PKG) and to cGMP dependent ion channels. The lowered intracellular calcium and activation of myosin light chain phosphatases results in inhibition of smooth muscle contractility and enhancement of erection. Multiple vasoactive agents, including vasoactive intestinal polypeptide (VIP), prostaglandin E1 (PGE1), forskolin, phosphodiesterase inhibitors and alpha adrenergic receptor antagonists affect smooth muscle contractility. Ultimately, all act by changes in intracellular calcium and modulation of specific smooth muscle myosin light chain kinases (MLCK) and myosin light chain phosphatases (MLCP). These enzymes rapidly change the state of myosin phosphorylation and result in either smooth muscle contraction (flaccidity) or relaxation (erection) [7].

Sexual dysfunction

Sexual problems are generally classified according to the four-phase model of sexual response [8]. These include: (i) sexual desire disorders (hypoactive sexual desire disorder [HSDD]); (ii) sexual arousal disorders (erectile dysfunction [ED]); and (iii) orgasmic disorders (premature or delayed ejaculation). Specific treatment for a sexual disorder depends on the results of a comprehensive biopsychosocial evaluation. Patients with primary (lifelong) or global (cross-situational) sexual dysfunction should be more intensively evaluated for organic etiologies, such as medical illness (e.g., renal insufficiency), medication side effects, endocrine factors (e.g., hypogonadism, hyperprolactinemia), and substance abuse.

Age-associated changes in male sexual response include 1) reduced libido, 2) reduced number and frequency of morning erections, 3) reduced penile sensitivity, 4) reduced arousal, including increased latency for and difficulty maintaining erection, 5) prolonged plateau phase, 6) reduced ejaculatory volume and force of expulsion, and 7) prolonged refractory period [9]. The age-associated decrease in testosterone level may be associated with a reduction in libido and mood, though this is controversial [10–12].

The most commonly encountered male sexual dysfunctions are HSDD, premature ejaculation (PE), and ED. These diagnoses require a great deal of clinical judgment, and do not include frequency or duration criteria. Male hypoactive sexual desire disorder is defined by the chronic absence (or deficiency in) sexual fantasies and desire for sexual activity [13, 14]. It has been estimated that the 1-year prevalence is 3% [15]. though other estimates are as high as 16% [16]. In men, HSDD is especially influenced the presence of other sexual dysfunctions (e.g., ED and PE), psychiatric illness, or medical conditions. It is particularly important to rule out hypogonadism and alcohol abuse, as these co-morbid conditions are common in men.

Premature ejaculation is defined as persistent or recurrent ejaculation with minimal sexual stimulation before, on, or shortly after penetration or before the person wishes it. An important component of the definition of PE is the patient's perception of lack of control over ejaculation. Generally, ejaculation occurs too quickly (less than one minute) and before, or soon after penetration into the vagina. Ejaculation occurring after vaginal penetration is less easily defined as premature. There are many couples for whom these definitions would be inadequate, and the most useful definition is probably an attitudinal one: ejaculation is premature if either partner perceives it to be. In making the diagnosis, clinicians should consider the circumstances of the problem, including the degree of associated distress in both the male and his partner.

Erectile dysfunction (ED) is defined as the inability to obtain and maintain an erection sufficient for satisfactory intercourse or other sexual expression. Approximately 10% of middle-aged men have complete erectile dysfunction, defined as the total inability to achieve or maintain erections sufficient for sexual performance. An additional 25% of men this age have intermittent erectile difficulties. The disorder is highly age-dependent, as the combined prevalence of moderate to complete erectile dysfunction rises from approximately 22% at age 40–49% by age 70 [17, 16, 15]. ED is associated with poor health: it is more common among men with diabetes, heart disease, hypertension, cigarette smoking, and hyperlipidemia [17]. There is a strong positive correlation between ED and depressive symptoms, impairment in social and occupational functioning, and reduced quality of life [18, 7]. Successful treatment of ED appears to reverse this morbidity [19].

Treatment of male sexual dysfunction

Hypoactive sexual desire disorder

Standard treatment for HSDD consists of individual or couples psychotherapy, with limited demonstrable treatment success [20]. Pharmacologic treatment with psychotropic agents such as yohimbine, dopamine agonists, stimulants, and/or antidepressants may stimulate mood changes or non-specific (i.e., placebo) responses, but have not been demonstrated to have specific therapeutic effects on HSDD. Testosterone replacement in hypogonadal men stimulates an increase in sexual desire and arousal [21].; in eugonadal men, the sexual effects of exogenous testosterone are indistinguishable from placebo [22].

Premature ejaculation

PE can be treated behaviorally, via the stop start or squeeze, or pharmacologically via SSRIs. The current method of choice in treating PE is behavior modification using the procedure described by Masters and Johnson [8]. This method is known as the pause method, and involves start and stop behavior therapy; combined with a frenulum squeeze procedure, it is called the pause–squeeze technique. The method involves direct stimulation of the male until premonitory sensations are experienced just prior to orgasm. All stimulation is stopped at this point. After repeated practices over 4–8 weeks, the male becomes more aware of these anticipatory sensations and is able to delay or control his ejaculation to a much greater degree. When practiced regularly by the male and his partner, the conditioning technique is often effective in the short term. However, it demands significant motivation on the part of the couple, and treatment efficacy is not always maintained over time. For example, a three-year follow-up study evaluating treatment and assessing long term benefits found that 75% of the patients showed no sign of lasting improvement over pretreatment baseline and there was no effect in 25% at the time of treatment termination [23]. Other disadvantages include the need for partner participation, and the time delay to achieve symptom remission [24].

Although SSRIs are not FDA-approved for PE, there is clear evidence in the literature of their effectiveness, and they are in widespread use. Since SSRIs had been known to delay ejaculation, the rationale, supported by case reports, was that a drug that delays ejaculation from the normal to impaired should delay ejaculation from premature to normal. In one study, men whose PE improved with clomipramine (CMI) received a comprehensive neurologic evaluation before and after treatment. The primary change demonstrated was a significant increase in genital sensory threshold [25]. The presumed mechanism is via 5HT2 receptor activation which is inhibitory to ejaculation – through ascending serotonergic projections to the medial preoptic area, and/or descending serotonergic pathways to the lumbosacral motor nuclei.

Subsequently, multiple controlled trials, using a variety of PE definitions, have supported the efficacy of serotonergic agents in this condition. In double-blind, placebo controlled trials using patient and partner assessments, CMI 25 mg and 50 mg used six hours prior to coitus [26, 27]., paroxetine 40 mg [28]., sertraline [29]., and fluoxetine 20–40 mg [30]. have been shown to produce significant ejaculatory delay – i.e., 4–10 min increased latency to ejaculation. Some authors have recommended initial daily use of one of these agents, followed by use as needed. In one study, rapid ejaculation resumed shortly after withdrawal of CMI therapy that had been administered for the previous eight weeks. To guard against this, patients should be encouraged to practice conditioning exercises along with use of the medication. No controlled studies have been performed to date on the combination of drug and non-drug therapies for this problem, although most authors currently recommend it [7].

Erectile dysfunction

The first-line option for ED is an oral phosphodiesterase type 5 inhibitor, such as sildenafil. This option can be used alone or in combination with sexual counseling or education. Brief sexual or couple's counseling is aimed at resolving specific conflicts, such as relationship distress, sexual performance concerns, and dysfunctional communication patterns. Advantages of sexual counseling include its noninvasiveness and relatively broad applicability. The major disadvantages are the lack of acceptability for many patients and uncertain efficacy.

Sildenafil is a competitive inhibitor of cyclic GMP-specific phosphodiesterase type 5 (PDE5), the predominant isozyme causing the breakdown of cyclic guanosine monophosphate (GMP) in the human corpus cavernosum. The mechanism of action of sildenafil is entirely peripheral, and includes direct effects on smooth muscle relaxation and vasodilation of the penile arterioles. Efficacy has been reported at 43–82%, depending primarily upon ED severity. The drug's side effects include headaches, flushing, dyspepsia, and nasal congestion, all of which are generally mild. A small percentage of men (less than 5%) may also experience alterations in color vision (blue hue), visual brightness or sensitivity, or blurred vision [31]. The effects (and side effects) are dose proportional, and the clinically effective dose range is between 25–100 mg. The only absolute contraindication to the administration of sildenafil is the concomitant use of nitrates or NO donors in any form including oral, sublingual, transdermal, inhalational or aerosol. This class of drugs may precipitate hypotension and syncope in the presence of sildenafil. Additionally, drugs, such as cimetidine or erythromycin that are metabolized by the same subclass of hepatic P450 enzymes may result in an increase in the sildenafil serum levels. However, these increases do not appear to be associated with an increase in adverse events

Currently, there are three additional oral agents in advanced stages of clinical development. These are vardenafil (Bayer–GlaxoSmithKline), tadalafil (Lilly/ICOS), and apomorphine (Abbott). Vardenafil and tadalafil are PDE5s that are more selective than sildenafil, and tadalafil is longer-acting ($T^{1}/_{2}$ = 17.5 h). Apomorphine is a centrally acting dopamine agonist. It has been available since 1869 and has been utilized parenterally as an emetic agent and as an anti-Parkinsonian drug. Both vardenafil and apomorphine have been approved in Europe.

Until the introduction of sildenafil, the primary non-surgical ED treatment with proven efficacy was penile self-injection therapy. This treatment consists of the injection of vasodilator medications into the penis using a small needle. After the initial test injections in the urologist's office, patients receive instructions in the self-injection technique. Once they have learned the proper technique and reached the satisfactory dose of the medication, patients receive medication and supplies for home injections. Follow-up is conducted through periodic visits to assure compliance and re-supply the medication. The most commonly utilized injectable medication is alprostadil, which is available as a ready-use-kit under the brand name Caverject. Injection therapy is effective in most cases of erectile dysfunction, regardless of etiology. It is contraindicated in men with a history of hypersensitivity to the drug employed, in those at risk for priapism (e.g., sickle cell disease, hypercoagulable states) and in men receiving monoamine oxidase inhibitors. Intracavernosal injections have been shown to be effective in about 70–80% of men who fail first-line therapy with sildenafil. However, many men and their partners find this method unacceptable: in six month follow-up studies of patients for whom injection therapy was effective, about half of the patients discontinued [32]. The current indication for self-injection therapy is as salvage for sildenafil non-responders or for those in whom sildenafil is contraindicated.

The use of vacuum constriction device (VCD) therapy is a well-established, non-invasive treatment that has recently been approved by the FDA for over the counter distribution. It provides a useful treatment alternative for patients for whom pharmacological therapies are contraindicated, or who do not desire other interventions. Vacuum constriction devices apply a negative pressure to the flaccid penis, thus drawing venous blood into the penis, which is then retained by the application of an elastic constriction band at the base of the penis. Efficacy rates of 60–80% have been reported in most studies. Like intracorporal injection therapy, VCD treatment is associated with a high rate of patient discontinuation. The adverse events occasionally associated with VCD therapy include penile pain, numbness, bruising, and delayed ejaculation.

Finally, surgical implantation of a penile prosthesis is appropriate in patients who fail non-surgical therapy for any reason, e.g., lack of efficacy, unacceptable side effects and/or dissatisfaction. Penile prostheses may also be indicated in patients with other conditions in addition to ED, such as Peyronie's disease and other penile deformities. The main advantages of penile prostheses are reliability, permanent availability, and freedom from medications and supplies. Disadvantages include need for surgery, infection and malfunction. After long years of development, penile prostheses have become a safe and effective treatment modality.

Depression and sexual dysfunction in men

Psychiatric illnesses can impact on sexual function in multiple ways. First, there can be direct neurobiological effects, as with the reversible loss of erectile function in a subset of depressed men. Second, there can be indirect effects through the disruption of social and interpersonal relationships. Finally, most of the drugs used in the treatment of psychiatric illnesses are associated with sexual dysfunction. In the setting of co-morbid sexual and psychiatric disorders, ascertaining which condition is primary is often difficult.

Models of co-morbidity

Although sexual dysfunction and depression are highly co-morbid, the causal relationship is unclear. There are five models, not mutually exclusive, which may be used to understand the coexistence of both conditions. First, the psychosocial distress that is invariably part of sexual dysfunction might stimulate the development of "secondary" depressive illness in vulnerable individuals. Second, sexual dysfunction can be a symptom of depression: MDD is associated with decreased libido, diminished erectile function, and decreased sexual activity. Third, a common factor (e.g., alcohol use, cardiovascular disease, or hypogonadism) might be etiologically related to both conditions. Fourth, antidepressant medication might lead to sexual dysfunction. Finally, depression and sexual dysfunction, both relatively common, might be serendipitously co-morbid and etiologically unrelated.

Impact of depression on sexual function

Loss of libido is considered a classic vegetative symptom of MDD and has played a prominent role in psychodynamic and other psychological formulations of depressive illness. Systematically collected data confirm that MDD is associated with decreased libido, diminished erectile function, and decreased sexual activity, though there are few studies of unmedicated depressed patients [33].

The most commonly used instrument for the measurement of male erectile response is a mechanical strain gauge that measures penile tumescence via an increase in penile circumference. The device is simple to use, and provides reliable measurement of nocturnal penile tumescence (NPT). Loss of NPT is a hallmark of ED that is presumed to be biologically mediated. Roose et al. [34]. first reported the observation that NPT was lost in two depressed men, and that this improved after depressive remission. This was then evaluated systematically by Steiger et al. [35]. who compared NPT in 25 men with an acute episode of depression to non-depressed control subjects. Although no statistically significant differences in NPT parameters were found between the depressed group and the control subjects, there was a complete lack of NPT in 4 (16%) depressed men that was reversed after antidepressant therapy. Thase et al. [36]. demonstrated a significant reduction in NPT time and decreased penile rigidity in 34 depressed men compared with non-depressed control subjects. NPT time was reduced by at least one standard deviation below the control mean in 40% of depressed men and was comparable to that in a group of 14 non-depressed men with a diagnosis of ED due to organic causes. These findings were confirmed in a repeat study with a new group of 51 men with MDD [33]. Together, these studies support the conclusion that erectile function is impaired in some depressed men, suggesting a neurophysiologic link between depression and ED.

The impact of sexual dysfunction on mood

The link between sexual function and mood has not been well characterized. There is, however, suggestive evidence of a strong bi-directional association from a population-based study of middle-aged men, multiple clinical studies performed at sexual dysfunction clinics, and from a recent ED treatment study. The Massachusetts Male Aging Study (MMAS) was a cross-sectional, community-based random-sample survey of health and aging in men aged 40 to 70 years (N = 1290). Based on the subjects' responses to nine questions that were highly correlated with biologic measures of erectile response, levels of ED were graded as nil (48%), minimal (17%), moderate (25%), or complete (10%) [17]. Depression and anger were highly correlated with ED. Using the Center for Epidemiologic Studies Depression Scale (CES-D) cutoff of 16 (suggesting clinically significant depressive symptoms), 41% of men with this degree of depressive symptomatology had moderate or complete ED – twice the level of non-depressed men. Maximal level of anger, as defined by Spielberger's anger scales, was associated with approximately 75% overall ED, double the ED prevalence among men who reported minimal anger [18].

In a large descriptive study, investigators carefully assessed co-morbid psychiatric conditions among men presenting to the Johns Hopkins Sexual Behaviors Consultation Unit from 1976–1979 (N = 199) and from 1984–1986 (N = 223) [37]. Overall, men with sexual dysfunction had high levels of depressive, somatic, and anxious symptoms and scored very high on measures of overall psychological distress (e.g., using one well-validated instrument that measures such distress, these men scored in the 92nd percentile of the normative population). Similarly, Strand et al. [38]. evaluated 121 men who presented to a sexual behaviors clinic with ED. Only 12% met criteria for a depressive disorder, but there was a high prevalence of dysphoria detected by the Brief Symptom Inventory.

Shabsigh et al. [39]. conducted a study of 100 men who presented to a urologic clinic with ED, benign prostatic hyperplasia (BPH), or both, and were screened for self-reported "depression", defined as above-threshold scores on two questionnaires. Men with ED (n = 66) were 2.6 times more likely to report "depression" than were men with BPH alone (n = 34). Moreover, among men who received ED treatment (with either penile intracavernosal injection or a vacuum device), all 15 patients in the non-depressed ED group continued treatment and were satisfied with the outcome, while only 7/18 (39%) of patients in the depressed ED group continued treatment. Thus, in this sample of men, depression was associated with ED, and with treatment discontinuation.

Finally, we conducted a placebo-controlled, parallel-group, double-blind study of 50–100 mg of sildenafil or placebo in 160 men with ED and co-morbid minor depression [19]. Patients were older than 18 years, were in stable heterosexual relationships, and had been experiencing ED for over six months. Minor depression was diagnosed with a structured clinical interview (SCID) by the presence of two to four DSM-IV major depressive episode criteria, with at least one being the A criterion (i.e., depressed mood or loss of interest or pleasure in most activities all day or every day for two weeks or longer) and by a score of 12 or greater on the 24-item Hamilton Rating Scale for Depression (HAM-D). Patients were randomized to sildenafil or placebo for 12 weeks. An "ED responder" was conservatively defined by a score of ≥ 22 on the erectile function domain (range 1–30) of the International Index of Erectile Function (IIEF) and affirmative responses to two questions regarding global improvement in erections and ability to have intercourse. The HAM-D and BDI scores were significantly reduced in ED responders (10.6 and 10.7, respectively) compared to ED non-responders (2.3 and 3.7, respectively, $p < 0.001$). Although the majority of men who were ED responders were in the sildenafil treatment group (83% *versus* 17%), it is interesting to note that even among the small number of ED responders treated with placebo, there were similar improvements in their depression scores. The results of this study suggest that depressed men who respond to ED treatment show robust improvement in depressive symptoms, and support the concept of depression secondary to medical illness.

Psychotropic medication and sexual dysfunction

Many prescription and non-prescription drugs with diverse pharmacologic properties have been reported to induce sexual dysfunction, particularly antihypertensives, antiulcer drugs, alcohol, sedative/hypnotics, mood stabilizers, antipsychotics, and antidepressants. Yet, most of these reports have been anecdotal, have focused only on men, and have not reliably identified the phase of sexual response that is impaired. More importantly, none has clearly distinguished the effects of the medication from the underlying disease. This is particularly important in assessing the sexual effects of psychotropic agents, since such effects should be considered in the context of behavioral changes in the underlying psychiatric condition associated with therapeutic actions of the drug. In addition, since the prevalence sexual dysfunction in the non-depressed general population is high [16]. linking sexual dysfunction to specific medications is difficult without population-based controls [40].

With the introduction and widespread use of SSRIs, the issue of medication-associated sexual dysfunction assumed greater clinical significance, particularly with respect to treatment compliance and quality of life concerns. In addition, in this context the neurobiological overlap between sexual function, depression, and monoamine physiology is particularly keen.

Antidepressants may cause sexual side effects in the drive phase (e.g., decreased libido – though this is difficult to distinguish from the decrease in sexual satisfaction associated with pervasive anhedonia); the arousal phase (e.g., ED – though the relation to pre-existing organic factors and to MDD complicates this association); and/or the release phase (e.g., delayed orgasm or anorgasmia). With serotonergic medications, orgasmic delay appears to be most common, followed by decreased libido, and then arousal difficulties (see [5]. for review). Painful ejaculation occurs in some men taking tricyclic antidepressants (TCAs). Estimates of SSRI-associated sexual side effects vary enormously – from less than 1% in postmarketing surveillance studies to greater than 50% in studies using systematic inquiries [5]. The best estimates are that about one-third of patients on SSRIs develop sexual dysfunction. Sexual dysfunction is reported somewhat less frequently with MAOIs, even less with TCAs, and rarely with nefazodone, bupropion, and mirtazepine.

Although low libido and ED are reported during SSRI treatment, the rates are not substantially different from base rates in the general population [16]. In contrast, a significant proportion of patients taking SSRIs report problems occurring during the orgasm phase (i.e., delayed ejaculation in men and anorgasmia in women); such problems are uncommon in the general population [5]. The presumed mechanism of orgasmic dysfunction is via 5-HT2 receptor activation which is inhibitory to ejaculation – through ascending serotonergic projections to the medial preoptic area and/or descending serotonergic pathways to the lumbosacral motor nuclei. Such activation apparently leads to an increase in genital sensory threshold, and the experience of genital "anesthesia". Some data suggest that SSRIs may affect sexual function via the reduced production of NO, and through noradrenergic actions [41]. PDE5s might have a pro-sexual effect via an impact on the NO system.

Strategies for treating SSRI-induced sexual dysfunction include: 1) decreasing the dose; 2) waiting for tolerance to develop; 3) switching to an antidepressant that does not have sexual side effects (e.g., nefazodone, bupropion, or mirtazepine); and/or 4) adding an "antidote" [42]. In anecdotal case reports, multiple "antidotes" have been reported, including the alpha2-adrenergic receptor antagonist, yohimbine [43].; the 5-HT2 receptor antagonists, nefazodone [44]. and cyproheptadine [45].; the 5-HT3 receptor antagonist, granisetron [46].; the dopamine agonist, amantadine [47].; bupropion [48].; buspirone [49].; psychostimulants [50]. and other agents, such as ginkgo biloba [51]. and sildenafil [42, 52]. In uncontrolled trials, "response" rates greater than 50% have been reported for many of these agents. All of these reports varied in dosing, follow-up, gender specificity, sexual dysfunction specificity, and methodology for determining sexual dysfunction remission. Indeed, none of these strategies has achieved widespread clinical acceptance.

There have been two placebo-controlled studies of agents for SSRI-associated sexual dysfunction. In one, amantadine and buspirone were not distinguishable from placebo in women with SSRI-associated sexual dysfunction [53]. In a second, 90 men who were taking SSRIs and reported treatment-emergent sexual dysfunction were randomized to receive either sildenafil or placebo. Sildenafil administration was associated with improvement in sexual dysfunction: 54.5% (24/44) of sildenafil compared with 4.4% (2/45) of placebo patients were much or very much improved (P < .001) [52]. However, all of the enrolled men had multiple sexual problems at baseline (mean 3.5 problems). Thus, beyond the established efficacy of sildenafil for ED, specific improvement in ejaculatory delay could not be distinguished from improvement in erectile function in this sample. This is particularly important because for the majority of men with SSRI-induced sexual dysfunction, persistent ejaculatory delay despite improved ED appears to be the norm.

Finally, in our experience, some men with SSRI-related ejaculatory delay were effectively treated with higher-than-usual doses of sildenafil. We treated 21 men (mean age = 56 years) who had MDD in remission and SSRI-associated ejaculatory delay with usual dose (25–100 mg) sildenafil. At baseline, 14/21 (67%) had co-morbid erectile dysfunction (ED). At the usual dose follow-up assessment, 12/14 achieved full ED remission, and 4/21 achieved ejaculatory delay remission. Sixteen patients with persistent ejaculatory delay were eligible for a high dose phase (100–200 mg sildenafil): five withdrew from the study, four took 150 mg maximum, and seven took 200 mg maximum. Of the 10 patients who had ejaculatory delay without significant ED and who chose to take high-dose sildenafil, nine reported a significant clinical improvement in ejaculatory delay and seven achieved full remission [54]. These open data suggest that anorgasmia may require a higher sildenafil dose for response than does ED.

Conclusion

The association between sexual dysfunction and mental illness is complex and bidirectional, though precise causal relationships remain unclear. Sexual dysfunction and the psychosocial distress that frequently accompanies it may precipitate the development of a mood disorder in vulnerable individuals, mood dysregulation might cause sexual dysfunction, or both conditions might coexist independently. Greater attention should be paid to sexual problems in depressed men. First, these men are at high risk for developing sexual problems either due to the mood disorder or to medications used in its treatment. Second, sexual problems can seriously interfere with quality of life in these patients. Finally, there are new options available for managing these problems.

References

1 Seidman SN, Rieder RO (1994) A review of sexual behavior in the United States. *Am J Psychiatry* 151: 330–341
2 Schiavi RC (1994) Effect of chronic disease and medication on sexual functioning. In: Rossi AS (ed): *Sexuality across the life course.* The University of Chicago Press, Chicago, 313–339
3 Meston CM, Frohlich PF (2000) The neurobiology of sexual function. *Arch Gen Psychiatry* 57: 1012–1030
4 Pfaus JG (1999) Neurobiology of sexual behavior. *Curr Opin Neurobiol* 9: 751–758
5 Rosen RC, Lane RM, Menza M (1999) Effects of SSRIs on sexual function: a critical review. *J Clin Psychopharmacol* 19: 67–85
6 Putnam SK, Du J, Sato S, Hull EM (2000) Testosterone restoration of copulatory behavior correlates with medial preoptic dopamine release in castrated male rats. *Horm Behav* 39: 216–224
7 Rosen RC, Goldstein I (2003) Sexual Dysfunction. In: David A. Warrell, Timothy M. Cox, John D. Firth, Edward J. Benz, Jr. (eds): *Oxford Textbook of Medicine.* Oxford University Press, Oxford,
8 Masters WH, Johnson VE (1970) *Human sexual inadequacy.* Churchill Livingstone, Edinburgh
9 Schiavi RC, Mandeli J, Schreiner-Engel P, Chambers A (1991) Aging, sleep disorders, and male sexual function. *Biol Psychiatry* 30: 15–24
10 McKinlay JB, Longcope C, Gray A (1989) The questionable physiologic and epidemiologic basis for a male climacteric syndrome: preliminary results from the Massachusetts Male Aging Study. *Maturitas* 11: 103–115
11 Morales A, Heaton JP, Carson CC (2000) Andropause: a misnomer for a true clinical entity. *J Urol* 163: 705–712
12 Seidman SN, Walsh BT (1999) Testosterone and depression in aging men. *Am J Geriatr Psychiatry* 7: 18–33
13 Segraves RT (1998) Definitions and classification of male sexual dysfunction. *Int J Impot Res* 10 Suppl 2: S54–S58
14 Schiavi RC, Segraves RT (1995) The biology of sexual function. *Psychiatr Clin North Am* 18: 7–23
15 Simons JS, Carey MP (2001) Prevalence of sexual dysfunctions: results from a decade of research. *Arch Sex Behav* 30: 177–219
16 Laumann EO, Paik A, Rosen RC (1999) Sexual dysfunction in the United States: prevalence and predictors. *JAMA* 281: 537–544
17 Feldman HA, Goldstein I, Hatzichristou DG, Krane RJ, McKinlay JB (1994) Impotence and its medical and psychosocial correlates: results of the Massachusetts Male Aging Study. *J Urol* 151: 54–61
18 Araujo AB, Durante R, Feldman HA, Goldstein I, McKinlay JB (1998) The relationship between depressive symptoms and male erectile dysfunction: Cross sectional results from the Massachusetts Male Aging Study. *Psychosom Med* 60: 458–465
19 Seidman SN, Roose SP, Menza MA, Shabsigh R, Rosen RC (2001) Treatment of erectile dysfunction in men with depressive symptoms: results of a placebo-controlled trial with sildenafil citrate. *Am J Psychiatry* 158: 1623–1630
20 Heiman JR (1998) Psychophysiological models of female sexual response. *Int J Impot Res* 10: S94–S97
21 Burris AS, Banks SM, Carter CS, Davidson JM, Sherins RJ (1992) A long-term, prospective study of the physiologic and behavioral effects of hormone replacement in untreated hypogonadal men. *J Andrology* 13: 297–304
22 Schiavi RC, White D, Mandeli J, Levine AC (1997) Effect of testosterone administration on sexual behavior and mood in men with erectile dysfunction. *Arch Sexual Behavior* 26: 231–241
23 St Lawrence JS, Madakasira S (1992) Evaluation and treatment of premature ejaculation: a critical review. *Int J Psychiatry Med* 22: 77–97
24 Althof SE (1998) Evidence based assessment of rapid ejaculation. *Int J Impot Res 10* (Suppl 2): S74–S76
25 Colpi GM, Fanciullacci F, Beretta G, Negri L, Zanollo A (1986) Evoked sacral potentials in subjects with true premature ejaculation. *Andrologia* 18: 583–586
26 Segraves RT, Saran A, Segraves K, Maguire E (1993) Clomipramine *versus* placebo in the treatment of premature ejaculation: a pilot study. *J Sex Marital Ther* 19: 198–200
27 Maniam P, Seftel AD, Corty EW, Rutchik SD, Hampel N, Althof SE (2001) Nocturnal penile

tumescence activity unchanged after long-term intracavernous injection therapy. *J Urol* 165: 830–832
28 Waldinger MD, Hengeveld MW, Zwinderman AH (1994) Paroxetine treatment of premature ejaculation: a double-blind, randomized, placebo-controlled study. *Am J Psychiatry* 151: 1377–1379
29 Mendels J, Camera A, Sikes C (1995) Sertraline treatment for premature ejaculation. *J Clin Psychopharmacol* 15: 341–346
30 Kara H, Aydin S, Yucel M, Agargun MY, Odabas O, Yilmaz Y (1996) The efficacy of fluoxetine in the treatment of premature ejaculation: a double-blind placebo controlled study. *J Urol* 156: 1631–1632
31 Seidman SN, Shabsigh R, Roose SP (1999) Pharmacologic treatment of sexual dysfunction. In: DL Dunner (ed): *Psychiatric Clinics of North America: Annual of Drug Therapy*. WB Saunders, Philadelphia, 21–33
32 Turner LA, Althof SE, Levine SB, Bodner DR, Kursh ED, Resnick MI (1992) Twelve-month comparison of two treatments for erectile dysfunction: self-injection *versus* external vacuum devices. *Urology* 39: 139–144
33 Nofzinger EA, Schwartz RM, Reynolds CF, Thase ME, Jennings JR, Frank E, Fasiczka AL, Garamoni GL, Kupfer DJ (1993) Correlation of nocturnal penile tumescence and daytime affect intensity in depressed men. *Psychiatry Res* 49: 139–150
34 Roose SP, Glassman AH, Walsh BT, Cullen K (1982) Reversible loss of nocturnal penile tumescence during depression: a preliminary report. *Neuropsychobiology* 8: 284–288
35 Steiger A, Holsboer F, Benkert O (1993) Studies of nocturnal penile tumescence and sleep electroencephalogram in patients with major depression and in normal controls. *Acta Psychiatr Scand* 87: 358–363
36 Thase ME, Reynolds CF, Jennings JR, Frank E, Howell JR, Houck PR, Berman S, Kupfer DJ (1988) Nocturnal penile tumescence is diminished in depressed men. *Biol Psychiatry* 24: 33–46
37 Fagan PJ, Schmidt CW, Wise TN, Derogatis LR (1988) Sexual dysfunction and dual psychiatric diagnoses. *Compr Psychiatry* 29: 278–284
38 Strand J, Wise TN, Fagan PJ, Schmidt CWJ (2002) Erectile dysfunction and depression: category or dimension? *J Sex Marital Ther* 28: 175–181
39 Shabsigh R, Klein LT, Seidman SN, Kaplan SA, Lehrhoff BJ, Ritter JS (1998) Increased incidence of depressive symptoms in men with erectile dysfunction. *Urology* 52: 848–852
40 Montgomery SA, Baldwin DS, Riley A (2002) Antidepressant medications: a review of the evidence for drug-induced sexual dysfunction. *J Affect Disord* 69: 119–140
41 Angulo J, Peiro C, Sanchez-Ferrer CF, Gabancho S, Cuevas P, Gupta S, Saenz de Tejada I (2001) Differential effects of serotonin reuptake inhibitors on erectile responses, NO-production, and neuronal NO synthase expression in rat corpus cavernosum tissue. *Br J Pharmacol* 134: 1190–1194
42 Nurnberg HG, Seidman SN, Gelenberg AJ, Fava M, Rosen R, Shabsigh R (2002) Depression, antidepressant therapies, and erectile dysfunction: clinical trials of sildenafil citrate (Viagra) in treated and untreated patients with depression. *Urology* 60: 58–66
43 Price J, Grunhaus LJ (1990) Treatment of clomipramine-induced anorgasmia with yohimbine: a case report. *J Clin Psychiatry* 51: 32–33
44 Reynolds RD (1997) Sertraline-induced anorgasmia treated with intermittent nefazodone [letter]. *J Clin Psychiatry* 58: 89
45 Lauerma H (1996) Successful treatment of citalopram-induced anorgasmia by cyproheptadine. *Acta Psychiatr Scand* 93: 69–70
46 Nelson EB, Keck PEJ, McElroy SL (1997) Resolution of fluoxetine-induced sexual dysfunction with the 5-HT3 antagonist granisetron [letter]. *J Clin Psychiatry* 58: 496–497
47 Balon R (1996) Intermittent amantadine for fluoxetine-induced anorgasmia. *J Sex Marital Ther* 22: 290–292
48 Labbate LA, Grimes JB, Hines A, Pollack MH (1997) Bupropion treatment of serotonin reuptake antidepressant-associated sexual dysfunction. *Ann Clin Psychiatry* 9: 241–245
49 Landen M, Eriksson E, Agren H, Fahlen T (1999) Effect of buspirone on sexual dysfunction in depressed patients treated with selective serotonin reuptake inhibitors. *J Clin Psychopharmacol* 19: 268–271
50 Roeloffs C, Bartlik B, Kaplan PM, Kocsis JH (1996) Methylphenidate and SSRI-Induced sexual side effects. *J Clin Psychiatry* 57: 548
51 Balon R (1999) Ginkgo biloba for antidepressant-induced sexual dysfunction? *J Sex Marital Ther* 25: 1–2

52 Nurnberg HG, Hensley PL, Gelenberg AJ, Fava M, Lauriello J, Paine S (2003) Treatment of antidepressant-associated sexual dysfunction with sildenafil: a randomized controlled trial. *JAMA* 289: 56–64
53 Michelson D, Bancroft J, Targum S, Kim Y, Tepner R (2000) Female sexual dysfunction associated with antidepressant administration: a randomized, placebo-controlled study of pharmacologic intervention. *Am J Psychiatry* 157: 239–243
54 Seidman SN, Pesce V, Roose SP (2000) High-dose Sildenafil Citrate for SSRI-associated Ejaculatory Delay: Open Clinical Trial. *J Clin Psychiatry* 2003; 64(6)

Potential role for the PDE5 inhibitor sildenafil in the treatment of female sexual dysfunction

Jennifer T. Anger and Jennifer R. Berman

The Female Sexual Medicine Center, UCLA Department of Urology, 924 Westwood Blvd., Suite 515, Los Angeles, California 90033, USA

Introduction

Although female sexual dysfunction has been recognized as a health problem that affects the well-being of a large population of both pre- and postmenopausal women, organic underlying causes that could be susceptible to pharmacologic or surgical treatment have only recently been considered [1]. In the last decade significant discoveries in the biochemistry, physiology and pharmacology of male sexual function have led to marked improvements in the clinical management of male sexual dysfunction [2–7]. In contrast, research in female sexual function and dysfunction has lagged significantly. This lag is attributed, in part, to the lack of basic science models of sexual responses in female animals, lack of understanding of female sexual anatomy and physiology, as well as lack of defined treatments. In addition, there has been a lack of funding for basic science research in the field.

Female sexual dysfunction is age-related, progressive, and highly prevalent, affecting 30–50% of American women [8, 9]. Studies comparing sexual dysfunction in couples have revealed that 40% of men had erectile or ejaculatory dysfunction whereas 63% of women had arousal or orgasmic dysfunction [10, 11]. In fact, approximately 20% of women report lubrication difficulties and the same percentage find sex not pleasurable [8]. The successful launch of sildenafil (Viagra®, Pfizer) for the treatment of male erectile dysfunction (ED) has inspired an increasing research effort in the pathophysiology and potential for pharmacologic treatment of female sexual dysfunction [12–15]. The question as to whether there were potential beneficial effects for Viagra in women soon followed.

Recent clinical and preclinical research studies have demonstrated that sildenafil enhances the female sexual arousal response. The American Foundation of Urological Diseases defines female sexual arousal disorder (FSAD) as "a persistent or recurring inability to attain or maintain sufficient sexual excitement that causes personal distress", which may be experienced as a lack of subjective excitement, genital lubrication/swelling, genital sensation or other somatic responses [16]. FSAD may occur independently of or concurrently with other distinct female sexual disorders, such as hypoactive sexual desire disorder (HSDD), female orgasmic disorder and/or dyspareunia. Although psychosocial factors clearly contribute to female sexuality and the sexual response, medical and physiological factors, including reduced vaginal/clitoral blood flow, altered hormonal environment, prior pelvic surgery (such as hysterectomy), vaginal injury from childbirth or use of certain medications, may be the primary basis for the disorder [17].

In human clitoral and vaginal tissue, pretreatment of smooth muscle strips with sildenafil enhances electrical field stimulation-induced relaxations [18], suggesting that nitric oxide (NO) is a key mediator of clitoral and vaginal smooth muscle relaxation. In the vagina NO appears to play a significant role in mediating vaginal smooth muscle relaxation [19]. Increased clitoral and vaginal vasocongestion during sexual arousal is in part mediated by the nitric oxide–cyclic guanosine monophosphate pathway. NO synthase and phosphodiesterase 5 (PDE5), the enzyme responsible for cyclic guanosine monophosphate catabolism, has been identified in human and animal clitoral cavernosal and vaginal tissue [20–22].

In men, cavernous nerves and endothelial cells release NO, stimulating formation of cyclic guanosine monophosphate (cGMP) by guanylate cyclase [23]. Sildenafil, an inhibitor of cGMP-specific phosphodiesterase type 5, selectively inhibits cGMP catabolism in cavernous smooth muscle tissue. In a similar fashion, clitoral and vaginal smooth muscle cells respond to exogenous nitric oxide by increased cGMP synthesis, which suggests that vaginal and clitoral cells express a functional nitric oxide/guanylyl cyclase/PDE5 pathway [24, 25]. cGMP is tightly regulated in these smooth muscle cells and the nitric oxide/cGMP pathway may be important in the regulation of clitoral and vaginal smooth muscle tone [25]. In fact, both *in vivo* animal models and clinical trials of women with FSAD have demonstrated enhanced genital blood flow and vaginal and clitoral engorgement with sildenafil treatment [26].

Pathophysiology of female sexual arousal disorder: vascular factors

Genital arousal in women is characterized by increased genital blood flow, vaginal lengthening, and vasocongestion of the vagina, vulva, and clitoris [27]. Increased vaginal blood flow during sexual arousal causes a transudate of fluid into the vagina, providing lubrication necessary for intercourse. Alterations to these processes can be associated with female sexual arousal disorder (FSAD). Women with FSAD have complaints of diminished vaginal lubrication, painful vaginal penetration, increased time for arousal, diminished vaginal or clitoral sensation, and difficulty in achieving orgasm [28]. These changes result, in part, from excessive deposition and disorganization of collagen fibers and loss of smooth muscle (tissue fibrosis) in the genital organs, a likely consequence of the same type of oxidative stress that causes vascular fibrosis in male ED.

The organic dysfunction underlying FSAD involves inadequate engorgement of the vagina, vulva and clitoris, as well as the loss of vaginal compliance and lubrication. FSAD is associated with aging, decreasing levels of estrogens and/or androgens, diabetes and vascular disease [12–14, 29–31]. Patients with atherosclerotic disease reportedly have reduced intensity of sexual excitement, arousal, as well as orgasm [32]. Diminished pelvic blood flow due to aortoiliac disease leads to vaginal wall and clitoral smooth muscle fibrosis resulting in symptoms of vaginal dryness and dyspareunia [33, 34]. In addition, traumatic injury to the iliohypogastric/pudendal arterial bed from pelvic fractures, blunt trauma, surgical disruption, or chronic perineal pressure from bicycle riding can result in diminished vaginal and clitoral blood flow and sexual dysfunction. Although a variety of psychological and medical disorders may result in decreased vaginal and clitoral engorgement, vascular insufficiency is an important cause of FSAD and should be considered in the evaluation of women with FSD.

Sildenafil in FSD: animal models

Animal models have demonstrated the importance of fibrosis as a cause of female sexual arousal disorder. Fibrosis results from collagen deposition by fibroblasts, myofibroblasts and smooth muscle cells themselves. The myofibroblast is a cell type that shares phenotypic features of both fibroblasts and smooth muscle cells [35]. Fibrosis of clitoral cavernosal tissue and the vagina is associated with aging, and experimentally has also been found to be associated with atherosclerosis-induced arterial insufficiency [10, 30]. In the female rabbit model, Park et al. induced atherosclerosis of the ilio-hypogastric-pudendal arterial bed by endothelial cell injury [30]. Pelvic nerve stimulation in these animals resulted in significantly lower vaginal blood flow, vaginal wall pressure and vaginal length compared to non-atherosclerotic controls. Diabetes mellitus, a known cause of vascular insufficiency, has also been shown to induce clitoral fibrosis in a rabbit model [13, 36]. In human clitoral tissue, there is a similar loss of corporal smooth muscle with replacement by fibrous connective tissue in association with atherosclerosis of clitoral cavernosal arteries [37].

Atherosclerotic changes that occur in clitoral vascular and trabecular smooth muscle may interfere with normal relaxation and dilation responses to sexual simulation. Using a rat model, other researchers have confirmed the same clitoral and vaginal hemodynamic changes reported by Park et al., specifically the increase in clitoral and vaginal blood flow after electrical field stimulation (EFS) [1, 15, 38]. Research by Min et al. showed that treatment with sildenafil caused a significant increase in genital blood flow and vaginal lubrication in intact and ovariectomized rabbits [39]. Estradiol treatment significantly increased both genital blood flow and vaginal lubrication above that observed in control animals.

Cultured rabbit vaginal and clitoral smooth muscle cells have been used to investigate the synthesis of second messenger cyclic nucleotides in response to vasodilators, and to determine the activity and kinetics of PDE5. Treatment with the NO donor, sodium nitroprusside, in the presence of sildenafil, enhanced cGMP synthesis and accumulation [1, 21]. In addition, a recent study in a rat model by Sato et al. [40] confirmed that intrathecal administration of sildenafil was associated with increased penile intracavernosal pressure. This suggests that sildenafil has both central and peripheral arousal enhancing effects and the NO/CGMP pathway likely plays a role in female sexual function as well [40].

Sildenafil in FSD: human studies

Pilot studies

Pilot studies of sildenafil in women with FSD have shown that sildenafil is well tolerated with a low incidence of side effects [23]. Minor side effects include headache, dizziness and dyspepsia. Some women may experience clitoral discomfort and "hypersensitivity". Such side effects usually do not cause patients to cease taking the medication. In a two-phase open-label pilot study using sildenafil to treat women with FSAD, the drug was similarly demonstrated to be safe and efficacious [40]. Following six weeks of home use of sildenafil, women demonstrated a significant improvement in subjective sexual function complaints. Post-stimulation physiologic measurements, including genital blood flow, vaginal pH and genital vibratory threshold measurements, also improved significantly compared to baseline. Patients were not stratified by hormonal status in this study [40].

Postmenopausal women

In a multicenter clinical trial of 201 spontaneously or surgically postmenopausal women with female sexual arousal disorder (FSAD), sildenafil was shown to significantly improve sexual function [40]. Each had undergone hysterectomy and/or was receiving estrogen and/or androgen replacement therapy. Hormonal requirements for the study included a free testosterone (0.9 pg/ml or greater) and estradiol (40 pg/ml or greater) corresponding to normal values for premenopausal women [43, 44]. Sildenafil (a 50 mg dose adjustable to 100 or 25 mg) was evaluated in a 12-week, double-blind, placebo controlled fashion. Primary endpoints were questions 2 (increased genital sensation during intercourse or stimulation) and 4 (increased satisfaction with intercourse and/or foreplay) from the Female Intervention Efficacy Index (FIEI). Secondary end points were the remaining questions from this index, the Sexual Function Questionnaire and sexual activity event log questions [40].

Sildenafil was effective and well tolerated in postmenopausal women with FSAD without concomitant hypoactive sexual desire disorder (HSDD or contributory emotional, relationship or historical abuse issues. Significant improvements in FIEI questions 2 ($p = 0.017$) and 4 ($p = 0.015$) were noted with sildenafil compared with placebo. For women with FSAD without concomitant HSDD, sildenafil was associated with significantly greater improvement in five of six FIEI items compared with placebo ($p < 0.02$). No significant improvements were shown for women with concomitant HSDD. Most adverse events were mild-to-moderate with headache, flushing, rhinitis, nausea and visual symptoms reported most frequently [40].

Premenopausal women

Sildenafil has recently been shown to improve sexual function in young, premenopausal women [41, 42]. In a double-blind, crossover, placebo-controlled trial of Viagra for premenopausal women affected by sexual arousal disorder, arousal and orgasm improved significantly with respect to placebo. The frequency of sexual fantasies, sexual intercourse and enjoyment improved in the women treated with sildenafil [41]. In another study by Caruso et al., in which healthy, asymptomatic premenopausal women underwent a double-blind, crossover, placebo-controlled trial of sildenafil, sildenafil significantly improved arousal, orgasm and enjoyment with respect to placebo [42]. This study suggests that sildenafil may act on different sexual pathways in healthy women, improving their sexual experience [42].

Women with neurologic disease

Sildenafil has been shown to improve sexual function in women with neurologic disease. In a double-blind, crossover design study of premenopausal women with spinal cord injuries, 19 women were randomly assigned to receive either sildenafil (50 mg) or placebo [45]. Physiologic and subjective measures of sexual response, heart rate, and blood pressure were recorded during baseline and sexual stimulation conditions. Significant increases in subjective arousal (SA) were observed with both drug ($P < 0.01$) and sexual stimulation conditions ($P < 0.001$), and a borderline significant ($P < 0.07$) effect of drug administration on vaginal pulse amplitude (VPA) was noted. Cardiovascular data showed modest increases in heart rate (±5 bpm) and mild decreases in blood pressure (±4 mm Hg) across all stimulation conditions, consistent with the peripheral vasodilatory mechanism of the drug. Sildenafil was well tolerated with no evidence of significant adverse events. Maximal responses occurred when sildenafil was combined with visual and manual sexual stimulation [45].

Although the prevalence of sexual dysfunction has been quoted as 72% among female patients with multiple sclerosis (MS) [46, 47], this aspect of patient wellbeing has been considerably neglected until recently. Symptoms include reduced sensation, reduced lubrication, difficulty reaching orgasm, dyspareunia and other types of pain disorders [47]. In a double-blind, placebo-controlled trial involving 217 men with MS and erectile failure, sildenafil significantly improved erectile function [48]. It is likely that sildenafil will similarly improve sexual function in women with MS.

FSD due to psychological issues

Women with sexual dysfunction due to psychological issues are less likely to respond to treatment with sildenafil. In a study evaluating the use of sildenafil over a six-week period at home, women with a history of unresolved childhood sexual abuse had a much lower rate of response to Viagra compared to those without a history of childhood sexual abuse [49].

Women on SSRIs

Sexual dysfunction has been observed in a substantial proportion of patients treated with all classes of antidepressants. In particular, SSRI use has been shown to be associated with sexual dysfunction [50]. Fava et al. evaluated the efficacy of sildenafil in a small sample of men and women with antidepressant-induced sexual dysfunction. 14 depressed outpatients were consecutively treated with oral sildenafil. Twelve of the 14 patients were treated with an SSRI and two with mirtazapine. All patients were prescribed oral sildenafil tablets at the initial dose of 50 mg q.d. p.r.n., with the possibility of increasing the dose to 100 mg q.d. p.r.n., if clinically indicated. A statistically significant improvement was found in all domains of sexual functioning, including libido, arousal, orgasm, sexual satisfaction, and (in males only) erectile function. 69% of patients reported themselves as much or very much improved. The findings of this study suggest that sildenafil may represent an efficacious approach to this population of patients.

The role of hormones in female sexual arousal and their effect: the NO/cGMP pathway

With menopause and its associated decline in circulating estrogen levels, the majority of women experience some degree of change in sexual function. Common sexual complaints include loss of desire, decreased frequency of sexual activity, painful intercourse, diminished sexual responsiveness, difficulty achieving orgasm, and decreased genital sensation. In 1966 Masters and Johnson first published their findings of the physiologic changes occurring in menopausal women that related to sexual function [51]. We have since learned that symptoms related to alterations in genital sensation and blood flow are in part secondary to declining estrogen levels, and that there is a direct correlation between the presence of sexual complaints and levels of estradiol below 50 pg/ml [52, 53]. Symptoms markedly decrease with estrogen replacement. Prior to sildenafil, the only available treatments for sexual complaints associated with menopause were hormone replacement therapy, commercial lubricants, psychotherapy and antidepressant medication.

Alterations in circulating estrogen levels associated with menopause contribute to age-associated changes in clitoral and vaginal smooth muscle. Fibrosis of the walls of the vaginal arteries results in a lack of lubrication during arousal [12]. In addition, fibrosis of the smooth muscle of the vagina itself, a known consequence of aging and estrogen deprivation, causes an inability of the vagina to stretch during intercourse. This may be one of the factors in dyspareunia, or pain during intercourse. Thinning of the vaginal epithelial layers, decreased vaginal submucosal microvasculature and diffuse clitoral cavernosal fibrosis have also been demonstrated in postmenopausal rabbit models after ovariectomy [54].

Estrogen also plays a role in regulating vaginal and clitoral nitric oxide synthase expression, the enzyme responsible for the production of nitric oxide (NO). Aging and surgical castration results in decreased vaginal and clitoral nitric oxide synthase expression and apoptosis of vaginal wall smooth muscle and mucosal epithelium. Estrogen replacement restores vaginal mucosal health, increases vaginal nitric oxide synthase expression, and decreases vaginal mucosal cell death [55]. These findings suggest that medications such as sildenafil mediate vascular and non-vascular smooth muscle via NO, and may have a potential role in the treatment of female sexual dysfunction, in particular that associated with sexual arousal disorder.

Testosterone supplementation is associated with increased well-being, energy, appetite and improved somatic and psychological scores in surgically menopausal women [56–58]. Surgically menopausal women given supraphysiologic doses of testosterone enanthate by intramuscular injection alone or in combination with estrogen experienced increased sexual desire, fantasies and arousal more often than those given placebo or estrogen alone [56]. Testosterone and estradiol implants also increased sexual activity, satisfaction, pleasure and frequency of orgasm more often than estrogen alone [57].

In another recent study [59], women who underwent hysterectomy and oophorectomy and were on estrogen replacement (at least 0.625 milligram of conjugated equine estrogen daily orally) were randomized to placebo, 150 or 300 micrograms of transdermal testosterone per day for 12 weeks each. The highest dose of testosterone resulted in a mean serum free testosterone slightly above the physiologic range and significantly increased scores for frequency of sexual activity and pleasure from orgasm. It also increased number of sexual fantasies, masturbation frequency and sense of positive well-being. Objectively, testosterone increased vaginal vasocongestion as measured by vaginal plethysmography during exposure to a potent visual stimulus in a small number of women with hypothalamic amenorrhea [59].

Significant improvements in arousal, orgasm, and frequency and enjoyment of sexual intercourse were previously reported in a double-blind, placebo controlled study evaluating sildenafil in premenopausal women with FSAD without HSDD [41]. In contrast, no significant improvement in sexual arousal was found in a randomized trial of sildenafil in a more heterogeneous population of postmenopausal women with FSAD, including only 46% with a primary diagnosis of FSAD, who were receiving estrogen but not androgen therapy [60].These findings suggest that adequate testosterone may be necessary for subjective improvement in sexual function and response with sildenafil treatment. Although studies have demonstrated that testosterone has a significant role in female sexual arousal, genital sensation orgasm and libido [59, 61, 62], determinations of normative testosterone values for women throughout the life cycle and appropriate therapeutic doses are still evolving [59].

Conclusions

Sildenafil has a role in the treatment of FSAD by improving blood flow to the genital tissue. When given in combination with long-term hormonal replacement therapy, sildenafil appears to have an even better effect on improvement in sexual function in the postmenopausal woman. In addition, recent studies have demonstrated an improvement in sexual function in premenopausal women with and without sexual arousal disorder [41, 42]. As the newer PDE5 inhibitors, such as vardenafil (Levitra®) and tadalafil (Cialis®), are proven to be efficacious in men, further studies will provide important information about their efficacy in women. In fact, vardenafil has already been shown to enhance clitoral and vaginal blood flow responses to pelvic nerve stimulation in female dogs [63].

Although there are similarities between men and women, the female sexual response is clearly multifaceted and distinct from that of the male [26]. Because medical/physiological factors and psychological/emotional factors may contribute to the development of sexual dysfunction, use of a combined medical and psychosexual evaluation for women seeking treatment for sexual dysfunction may result in more efficient identification of potential causes and selection of appropriate treatment.

References

1 Munarriz R, Kim SW, Kim NN, Traish A, Goldstein I (1992) A review of the physiology and pharmacology of peripheral (vaginal and clitoral) female genital arousal in the animal model. *J Urol* 170: S40
2 Lue TF (1995) A patient's goal-directed approach to erectile dysfunction and Peyronie's disease. *Can J Urol* 2: 13–17
3 Lin CS, Lin G, Lue TF (2003) Isolation of two isoforms of phosphodiesterase 5 from rat penis. *Int*

J Impot Res 15: 129–136
4 Lue TF (2003) Phosphodiesterases as therapeutic targets. *Urology* 61: 685–691
5 Bivalacqua TJ, Usta MF, Champion HC, Adams D, Namara DB, Abdel-Mageed AB, Kadowitz PJ, Hellstrom WJ (2003) Gene transfer of endothelial nitric oxide synthase partially restores nitric oxide synthesis and erectile function in streptozotocin diabetic rats. *J Urol* 169: 1911–1917
6 Sikka SC, Hellstrom WJ (2002) Role of oxidative stress and antioxidants in Peyronie's disease. *Int J Impot Res* 14: 353–360
7 Hellstrom WJ, Gittelman M, Karlin G, Segerson T, Thibonnier M, Taylor T, Padma-Nathan H; Vardenafil Study Group (2003) Sustained efficacy and tolerability of vardenafil, a highly potent selective phosphodiesterase type 5 inhibitor, in men with erectile dysfunction: results of a randomized, double-blind, 26-week placebo-controlled pivotal trial. *Urology* 61: 8–14
8 Laumann EO, Michael RT, Gagnon JH (1994) A political history of the national sex survey of adults. *Fam Plann Perspect* 26: 34–38
9 Berman JR, Bassuk J (2002) Physiology and pathophysiology of female sexual function and dysfunction. *World J Urol* 20: 111–118
10 Park K, Goldstein I, Andry C, Siroky MB, Krane RJ, Azadzoi KM (1997) Vasculogenic female sexual dysfunction: The hemodynamic basis for vaginal engorgement insufficiency and clitoral erectile insufficiency. *Int J Impot Res* 9: 27–37
11 Frank E, Anderson C, Rubinstein D (1978) Frequency of sexual dysfunction in 'normal' couples. *N Eng J Med* 299: 111–115
12 Myers LS, Morokoff PJ (1986) Physiological and subjective sexual arousal in pre- and postmenopausal women taking replacement therapy. *Psychophysiology* 23: 283–292
13 Park K, Ahn K, Chang JS, Lee SE, Ryu SB, Park YI (2002) Diabetes induce alteration of clitoral hemodynamics and structure in the rabbit. *J Urol* 168: 1269–1272
14 Sherwin BB, Gelfand MM, Brender W (1985) Androgen enhances sexual motivation in females. A prospective, cross over study of sex steroid administration in surgical menopause. *Psychosom Med* 47: 339–351
15 Giuliano F, Allard J, Compagnie S, Alexandre L, Droupy S, Bernabe J (2001) Vaginal physiological changes in a model of sexual arousal in anesthetized rats. *Am J Physiol Regul Integr Comp Physiol* 281: R140–R149
16 Basson R, Berman J, Burnett A, Derogatis L, Ferguson D, Fourcroy J, Goldstein I, Graziottin A, Heiman J, Laan E et al. (2000) Report of the International Consensus Development Conference on Female Sexual Dysfunction: definitions and classifications. *J Urol* 163: 888–893
17 Goldstein I (2000) Female sexual arousal disorder: new insights. *Int J Impot Res* 12: S152–S157
18 Vemulapalli S, Kurowski S (2000) Sildenafil relaxes rabbit clitoral corpus cavernosum. *Life Sci* 67: 23–29
19 Ziessen T, Moncada S, Cellek S (2002) Characterization of the non-nitrergic NANC relaxation responses in the rabbit vaginal wall. *Br J Pharmacol* 135: 546–554
20 Park K, Moreland RB, Goldstein I, Atala A, Traish A (1998) Sildenafil inhibits phosphodiesterase type 5 in human clitoral corpus cavernosum smooth muscle. *Biochem Biophys Res Commun* 249: 612–617
21 Traish A, Moreland RB, Huang YH, Kim NN, Berman J, Goldstein I (1999) Development of human and rabbit vaginal smooth muscle cell cultures: effects of vasoactive agents on intracellular levels of cyclic nucleotides. *Mol Cell Biol Res Commun* 2: 131–137
22 D'Amati G, di Gioia CR, Bologna M, Giordano D, Giorgi M, Dolci S, Jannini EA (2002) Type 5 phosphodiesterase expression in the human vagina. *Urology* 60: 191–195
23 Kaplan SA, Reis RB, Kohn IJ, Ikeguchi EF, Laor E, Te AE, Martins AC (1999). Safety and efficacy of sildenafil in postmenopausal women with sexual dysfunction. *Urology* 53: 481–486
24 Traish AM, Moreland RB, Huang YH, Kim NN, Berman J, Goldstein I (1999) Development of human and rabbit vaginal smooth muscle cultures: effects of vasoactive agents on intracellular levels of cyclic nucleotides. *Mol Cell Biol Res Commun* 2: 131–137
25 Shabsigh R (2002) Female Sexual Function and Dysfunction. In: PC Walsh, AB Retik, ED Vaughan, Jr., AJ Wein, LR Kavoussi, AC Novick, AW Partin, CA Peters (eds): *Campbell's Urology*, 8th Edition. Volume 2. WB Saunders, Philadelphia, 1710–1733
26 Berman JR, Berman LA, Lin H, Flaherty E, Lahey N, Goldstein I, Cantey-Kiser J (2001) Effect of sildenafil on subjective and physiologic parameters of the female sexual response in women with sexual arousal disorder. *J Sex Marital Ther* 27: 411–420
27 Traish AM, Kim N, Min K, Munarriz R, Goldstein I (2002) Role of androgens in female genital

sexual arousal: receptor expression, structure, and function. *Fertil Steril* 77 (Suppl 4): S11–S18
28 Traish AM, Kim NN, Munarriz R, Moreland R, Goldstein I (2002) Biochemical and physiological mechanisms of female genital arousal. *Arch Sex Behav* 31: 393–400
29 Goldstein I, Berman JR (1998) Vasculogenic female sexual dysfunction: vaginal engorgement and clitoral erectile insufficiency syndromes. *Int J Impot Res* 10 (Suppl 2): S84–S90
30 Park K, Tarcan T, Goldstein I, Siroky MB, Krane RJ, Azadzoi KM (2000) Atherosclerosis-induced chronic arterial insufficiency causes clitoral cavernosal fibrosis in the rabbit. *Int J Impot Res* 12: 111–116
31 Traish AM, Kim N, Min K, Munarriz R, Goldstein I (2002): Role of *androgens in female genital sexual arousal: receptor expression, structure, and function. Fertil* Steril 77 (Suppl 4): S11–S18
32 Mooradian AD, Greiff V (1990) Sexuality in older women. *Arch Intern Med* 150: 1033–1038
33 Berman JR, Berman L, Goldstein I (1999) Female sexual dysfunction: incidence, pathophysiology, evaluation, and treatment options. *Urology* 54: 385–391
34 Goldstein I, Berman JR (1998) Vasculogenic female sexual dysfunction: vaginal engorgement and clitoral erectile insufficiency syndromes. *Int J Impot Res* 10 (Suppl 2): S84–90; discussion S98–101
35 Vernet D, Ferrini MG, Valente E, Magee TR, Bou-Gharios G, Rajfer J, Gonzalez-Cadavid NF (2002) Effect of nitric oxide on fibroblast differentiation into myofibroblasts in cell cultures from the Peyronie's fibrotic plaque and in its rat model *in vivo*. *Nitric Oxide* 7: 262–276
36 Valente EG, Ferrini MG, Vernet D, Qian A, Rajfer J, Gonzalez-Cadavid NF (2003) L-arginine and phosphodiesterase (PDE) inhibitors counteract fibrosis in the Peyronie's fibrotic plaque and related fibroblast cultures. *Nitric Oxide* 9: 229–244
37 Tarcan T, Park K, Goldstein I, Maio G, Fassina A, Krane RJ, Azadzoi KM (1999) Histomorphometric analysis of age-related structural changes in human clitoral cavernosal tissue. *J Urol* 161: 940–944
38 Vachon P, Simmerman N, Zahran AR, Carrier S (2000) Increases in clitoral and vaginal blood flow following clitoral and pelvic plexus nerve stimulations in the female rat. *Int J Impot Res* 12: 53–57
39 Min K, Munarriz R, Kim NN, Goldstein I, Traish A (2002) Effects of ovariectomy and estrogen and androgen treatment on sildenafil-mediated changes in female genital blood flow and vaginal lubrication in the animal model. *Am J Obstet Gynecol* 187: 1370–1376
40 Berman JR, Berman LA, Toler SM, Gill J, Haughie S; for the Sildenafil Study Group (2003) Safety and Efficacy of Sildenafil Citrate for the Treatment of Female Sexual Arousal Disorder: A Double-Blind, Placebo Controlled Study. *J Urol* 170: 2333–2338
41 Caruso S, Intelisano G, Lupo L, Agnello C (2001) Premenopausal women affected by sexual arousal disorder treated with sildenafil: a double-blind, cross-over, placebo-controlled study. *BJOG* 108: 623–628
42 Caruso S, Intelisano G, Farina M, Di Mari L, Agnello C (2004) The function of sildenafil on female sexual pathways: a double-blind, cross-over, placebo-controlled study. Eur J Obstet Gynecol, *in press*
43 Guay A, Munarriz R, Jacobson MA, Talakoub L, Goldstein I, Traish A et al. (2002) Androgen values in premenopausal women without sexual dysfunction. Presented at International Society for the Study of Women's Sexual Health, Vancouver, British Columbia, Canada, October 10–13
44 Korenman SG (1982) Menopausal endocrinology and management. *Arch Intern Med* 142: 1131–1136
45 Sipski ML, Rosen RC, Alexander CJ, Hamer RM (2000) Sildenafil effects on sexual and cardiovascular responses in women with spinal cord injury. *Urology* 55: 812–815
46 Lillius H, Valtonene C, Wilkstrom J (1976) Sexual problems in patients suffering from multiple sclerosis. *J Chron Dis* 19: 643–647
47 DasGupta R, Fowler CJ (2003) Bladder, Bowel and Sexual Dysfunction in Multiple Sclerosis: Management Strategies. *Drugs* 63: 153–166
48 Fowler C, Miller J, Sharief M (1999) Viagra (Sildenafil Citrate) for the treatment of erectile dysfunction in men with multiple sclerosis. *Ann Neurol* 46: 497
49 Berman LA, Berman JR, Bruck D, Pawar RV, Goldstein I (2001) Pharmacotherapy or psychotherapy?: effective treatment for FSD related to unresolved childhood sexual abuse. *J Sex Marital Ther* 27: 421–425
50 Fava M, Rankin MA, Alpert JE, Nierenberg AA, Worthington JJ (1998) An open trial of oral sildenafil in antidepressant-induced sexual dysfunction. *Psychother Psychosom* 67: 328–331
51 Masters WH, Johnson VE (1966) Human Sexual Response. Little Brown and Co., Boston, MA

52 Sarrell P (1990) Sexuality and menopause. *Obstet Gynecol* 75 (4 Suppl): 26S-30S
53 Sarrel PM (1998) Ovarian hormones and vaginal blood flow: using laser Doppler velocimetry to measure effects in a clinical trial of post-menopausal women. *Int J Impot Res* 10 (Suppl 2): S91-93; discussion S98-101
54 Park K, Ahn K, Lee S, Ryu S, Park Y, Azadzoi KM (2001): Decreased circulating levels of estrogen alter vaginal and clitoral blood flow and structure in the rabbit. *Int J Impot Res* 13: 116-124
55 Berman JR, McCarthy MM, Kyprianou N (1998) Effect of estrogen withdrawal on nitric oxide synthase expression and apoptosis in the rat vagina. *Urology* 51: 650-656
56 Sherwin BB, Gelfand MM, Brender W (1985) Androgen enhances sexual motivation in females: a prospective, crossover study of sex steroid administration in the surgical menopause. *Psychosom Med* 47: 339-351
57 Davis SR, McCloud P, Strauss BJ, Burger H (1995) Testosterone enhances estradiol's effects on postmenopausal bone density and sexuality. *Maturitas* 21: 227-236
58 Davis S (1999) Androgen replacement in women: a commentary. *J Clin Endocrinol MeTab.* 84: 1886-1891
59 Shifren JL, Braunstein GD, Simon JA, Casson PR, Buster JE, Redmond GP, Burki RE, Ginsburg ES, Rosen RC, Leiblum SR et al. (2000) Transdermal testosterone treatment in women with impaired sexual function after oophorectomy. *N Engl J Med* 343: 682-688
60 Basson R, McInnes R, Smith MD, Hodgson G, Koppiker N (2002) Efficacy and safety of sildenafil citrate in women with sexual dysfunction associated with female sexual arousal disorder. *J Womens Health Gend Based Med* 11: 367-377
61 Sherwin BB, Gelfand MM (1987) The role of androgen in the maintenance of sexual functioning in oophorectomized women. *Psychosom Med* 49: 397-409
62 Marin R, Escrig A, Abreu P, Mas M (1999) Androgen-dependent nitric oxide release in rat penis correlates with levels of constitutive nitric oxide synthase isoenzymes. *Biol Reprod* 61: 1012-1016
63 Angulo J, Cuevas P, Cuevas B, Bischoff E, Saenz de Tejada I (2003) Vardenafil enhances clitoral and vaginal blood flow responses to pelvic nerve stimulation in female dogs. *Int J Impot Res* 15: 137-141

Molecular processing of sildenafil in endothelial function: potential applications in cardiovascular diseases

Shadwan F. Alsafwah and Stuart D. Katz

Department of Internal Medicine, Section of Cardiovascular Medicine, Yale University School of Medicine, 135 College Street, Suite 301, New Haven, CT 06510, USA

Nitric oxide signaling in vascular smooth muscle

Nitric oxide (NO), released by vascular endothelial cells in response to hormonal and physiochemical stimuli, induces vasorelaxation by increasing production of the second messenger cyclic guanosine monophosphate (cGMP) via the activation of the heme-containing protein soluble guanylate cyclase in vascular smooth muscle [2–4]. In vascular smooth muscle, cGMP mediates its vasorelaxation effects by activation of cGMP-dependent protein kinase (PKG), which phosphorylates target downstream proteins that regulate ion channel function and intracellular calcium concentration [5, 6]. Tissue cGMP levels are determined by the balance between the activities of the guanylate cyclases that catalyze formation of cGMP (soluble and membrane-bound forms) and the cyclic nucleotide phosphodiesterases that catalyze breakdown of cGMP [7, 8]. Phosphodiesterase type 5 (PDE5) is a key regulatory enzyme for cGMP signaling in vascular smooth muscle. PDE5 is a homodimer with a catalytic domain with two zinc-binding motifs, two allosteric binding sites for cGMP, and a regulatory phosphorylation site at serine 92 [9]. Binding of cGMP to the catalytic site on the PDE5 molecule results in cleavage to guanosine monophosphate and increased binding to the allosteric regulatory sites [8, 10]. The role of the allosteric binding sites has not been fully characterized, but cGMP binding to these sites has been reported to increase catalytic activity [11, 12]. PKG phosphorylation of the serine 92 regulatory site increases PDE5 activity to create a negative feedback loop that limits the magnitude and duration of response to cGMP stimulating signals [11, 13, 14]. Potentiation of cGMP signaling by pharmacological inhibition of phosphodiesterase type 5 (PDE5) is a proven therapeutic strategy in patients with erectile dysfunction [9, 15]. This review will discuss the potential role of inhibition of PDE5 as a therapeutic strategy to augment endothelium-dependent NO-cGMP signaling in the coronary, pulmonary, and systemic vasculature in patients with cardiovascular diseases.

Characterization of endothelial dysfunction in cardiovascular diseases

The obligatory role of the vascular endothelium in mediating the vasodilatory response to acetylcholine in rat aorta was first reported by Furchgott and colleagues in 1980 [16]. Subsequent investigations determined that the "endothelium-derived relaxing factor" described by Furchgott was chemically indistinguishable from nitric oxide [3, 17, 18]. Nitric oxide is synthesized in endothelial cells from the guanidine nitrogens of the amino acid L-arginine by a constitutively expressed heme-containing reductase enzyme endothelial nitric oxide synthase (eNOS also described as NOS3) [4]. The activity of eNOS is regulated by numerous factors including substrate availability, cellular redox state, availability of calcium, calmodulin, and required cofactors NADPH and tetrahydrobiopterin, and post-translational modification and translocation [19, 20]. Endothelial dysfunction was first described in clinical studies as a paradoxical arterial vasoconstriction response to intra-coronary infusion of the endothelium-dependent vasodilator, acetylcholine, in patients with coronary artery disease in 1986 [21]. Numerous subsequent studies have demonstrated impaired endothelium-dependent vasodilation, either in response to regional intra-arterial infusions or shear stress-induced, flow-mediated dilation, in the coronary and systemic circulations in patient populations with established coronary artery disease and subjects with traditional and non-traditional risk factors for coronary artery disease [22, 23]. Endothelium-dependent vasodilation has also been shown to decrease acutely in normal subjects in response to rapid rises in serum lipids after a fatty meal and after rapid rises in plasma homocysteine levels after ingestion of methionine [24, 25]. The etiology of endothelial dysfunction in these clinical and experimental settings has not been fully characterized and is likely multifactorial (Tab. 1) [26]. Reduced bioavailability of NO, due to decreased activity of the L-arginine–NO biosynthetic pathway and/or increased degradation of NO by reactive oxygen species are thought to be contributory factors [27–29]. An important role of reactive oxygen species is supported by clinical studies that demonstrated improved endothelial function in response to antioxidant vitamins, antioxidant iron chelators, and supplementation of tetrahydrobiopterin, an essential co-factor for eNOS [24, 30, 31].

Impaired endothelium-dependent vasodilation is a convenient biomarker for assessment of endothelial dysfunction in clinical studies that is thought to be associated with proinflammatory and prothrombotic changes in the endothelial cell phenotype [22, 32]. Abnormal endothelium-dependent vasomotion is associated with increased expression of cell surface molecules that increase adhesion and infiltration of macrophages and lymphocytes, increased platelet adhesion and aggregation, and increased activity of plasminogen activator inhibitor-1 [22, 32]. Increased inflammation and prothrombotic activity are thought to play an important role in the pathogenesis of atherosclerosis progression and acute coronary syndromes [22, 32]. Recent studies have demonstrated that impaired endothelium-dependent vasodilation

Table 1. Potential causes of impaired endothelium-dependent vasodilation in heart failure

1. Abnormalities of endothelial cell function
    A. Decreased production of nitric oxide
        i. Decreased bioavailability of eNOS substrate L-arginine
            a. Decreased L-arginine transport via CAT-1
            b. Increased production of asymetric dimethylarginine
        ii. Decreased expression/synthesis of endothelial nitric oxide synthase
            a. Decreased transcription message
            b. Shortened transcript half-life
            c. Decreased translation
        iii. Altered regulation of endothelial nitric oxide synthase
            a. Post-translational modification
            b. Dysregulation of chaperone proteins (caveolin, heat shock proteins)
            c. Dysregulation of protein phosphorylation
            d. Dysregulation of calcium, redox state or other co-factor metabolism
        iv. Altered stimuli for eNOS activation
            a. Decreased shear stress signal
            b. Altered neurohormonal milieu
            c. Dysfunctional receptors
    B. Increased degradation of nitric oxide
        i. Increased diffusion distance
        ii. Increased in production of reactive oxygen species
            a. Membrane bound NAD(P)H oxidase
            b. Xanthine oxidase
            c. Mitochondrial sources
            d. Iron-mediated Fenton chemistry
    C. Increased production of counter-regulatory vasoconstrictors

2. Abnormalites of vascular smooth muscle cell function
    A. Decreased bioavailability of cGMP
        i. Decreased expression and/or activity of soluble guanylate cyclase
        ii. Increased expression and/or activity of phosphodiesterase type 5
    B. Downstream abnormalities in PKG signaling pathway
    C. Increased action of counter-regulatory vasoconstrictors

in the coronary and peripheral circulations is associated with increased risk of subsequent major cardiovascular events in patients with established coronary artery disease, patients with peripheral vascular disease, and patients with hypertension [33].

The pathophysiological role of endothelial dysfunction in chronic heart failure (CHF) is more complex than that of other cardiovascular disease populations [34]. Exercise intolerance in CHF is attributable to reduction in maximal cardiac output reserve and peripheral vascular changes that alter regional distribution of cardiac output [34, 35]. Decreased metabolic vasodilatory reserve in the skeletal muscle circulation during exercise reduces peripheral oxygen delivery and peak aerobic capacity in patients with CHF [36–40]. Flow-mediated nitric oxide (NO)-mediated vasodilation during exercise is impaired in the skeletal muscle circulation of rats with post-myocardial infarction heart failure [41], and in the forearm circulation of patients with CHF [42]. In patients with CHF, maximal exercise capacity and New York Heart Association functional class are significantly associated with the severity of impairment of NO-mediated vasodilation [42–45].

Decreased endothelium-dependent vasodilation in patients with CHF is most certainly multifactorial and can be partly attributed to decreased shear stress signal in arterial circulations due to reduced cardiac output and reduced activity of the L-arginine–NO metabolic pathway [27, 40]. Increased degradation of NO by reactive oxygen species may also contribute to reduced NO-mediated vasodilation in heart failure [28, 46]. In addition, impaired vasodilation in response to endothelium-derived NO in patients with CHF is partly attributable to hyporesponsiveness of cGMP-dependent vasorelaxation in vascular smooth muscle. Most published reports indicate that regional vasodilatory responses to intra-arterial administration of nitrosovasodilators are significantly decreased in patients with CHF when compared with normal subjects [38, 47, 48]. Vasodilatory responses to intra-arterial administration of natriuretic peptides, which mediate vasodilation through activation of particulate guanylate cyclase in vascular smooth muscle, are also attenuated in patients with CHF when compared with normal subjects [49]. In contrast, the vasodilation responses cyclic adenosine monophosphate (cAMP) dependent vasodilators, isoproterenol and phenotolamine, did not differ in patients with CHF when compared with normal subjects [50, 51]. The sum of these observations suggests that a specific defect in cGMP-mediated vasodilation is present in patients with CHF.

Impaired cGMP-mediated vasodilation in heart failure may be due to decreased cGMP synthesis by soluble and/or particulate forms of guanylate cyclase, increased cGMP degradation by phosphodiesterase, or downstream abnormalities in the cGMP signal transduction pathway [5]. In isolated glomeruli from dogs with heart failure induced by rapid ventricular pacing, cGMP accumulation in response to atrial natriuretic peptide and sodium nitroprusside is decreased when compared with normal controls and is associated with increased phosphodiesterase hydrolyzing activity of cGMP [52]. In cultured rat mesangial cells, angiotensin II attenuates the effects of atrial natriuretic peptide on cGMP accumulation in part by increasing phosphodiesterase hydrolyzing activity [53]. These experimental observations suggest that increased phosphodiesterase activity may contribute to decreased cGMP-mediated vasodilation in patients with CHF. Accordingly, selective inhibition of PDE5, the most abundant isoform of cGMP-hydrolyzing phosphodiesterases in vascular smooth muscle, is a potential therapeutic strategy to enhance endothelium-dependent NO-mediated dilation in patients with CHF.

Potential therapeutic role of PDE5 inhibition in CV disease

The 11 isoforms of PDE described to date, each the product of separate genes, are classified according the isoform tissue distribution and substrate specificity (cAMP or cGMP) [7, 8, 54]. Phosphodiesterase type 5 (PDE5) is the predominant isoform that regulates cGMP content in vascular smooth muscle [7, 8, 55]. PDE5 has also been reported to be present in skeletal and visceral muscles and platelets, but has not been found in human cardiac myocytes [7, 56–58]. The different distribution of PDE isoforms in various tissues as well as the selectivity of different pharmacological agents constitute the basis for the tissue-specific effects of PDE inhibitors [56].

Sildenafil is a highly selective inhibitor of PDE5 approved for the treatment of erectile dysfunction in 1998 [8, 9, 59]. Sildenafil is structurally similar to the native PDE5 substrate, cGMP, and competes with cGMP for binding at the catalytic site (but not the allosteric sites) on the PDE5 molecule [8, 10]. Site directed mutagenesis indicates that binding of sildenafil at the catalytic domain in increased in the presence of cGMP [8, 12]. The clinical benefits of sildenafil in erectile dysfunction are attributable to enhanced cGMP-dependent effects related to NO signaling in corpus cavernosum (a specialized type of vascular smooth muscle) in the penile circulation [59, 60]. Vardenafil is a structurally similar PDE5 inhibitor also approved for treatment of erectile dysfunction with PDE5 selectivity comparable to that of sildenafil and greater potency on a milligram basis [61–63]. Both sildenafil and vardenafil have some inhibitory effects on PDE6 (which is mainly distributed in the photoreceptor cells of the retina) with infrequent reports of visual side effects in clinical trials [9]. Tadalafil, the third selective inhibitor of PDE5 approved for the treatment of erectile dysfunction, is structurally distinct with a much longer half-life and some inhibitory effects on PDE11, a recently described isoform that is widely distributed in cardiac and skeletal muscle, prostate, testes, kidney, liver, and pituitary [9, 64, 65]. The clinical significance of PDE11 inhibition effects of tadalafil is not known.

Based on the available evidence presented in the above discussion, it is possible that impaired NO-mediated vasodilation in response to endothelium-derived NO or organic nitrate-derived NO is at least partially attributable to the increased activity of PDE5 in vascular smooth muscle. The effects of acute PDE5 inhibition on endothelial function have been reported in experimental and clinical studies of coronary artery disease, CHF, diabetes mellitus and pulmonary hypertension.

Coronary artery disease

In chronically-instrumented conscious dogs with surgically-induced coronary stenosis, acute administration of sildenafil 2 mg/kg was associated with improved myocardial blood flow in the ischemic region. The specific role of nitric oxide was not determined in this experimental model [66]. Halcox et al. studied regional vascular responses to infusion of acetylcholine in 15 patients with established coronary artery disease before and 45 min after administration of sildenafil 100 mg [67]. Coronary arterial segments that contracted after application of acetylcholine at baseline, indicating endothelial dysfunction, dilated after the combined administration of sildenafil with acetylcholine, suggesting that PDE5 inhibition could acutely enhance endothelial function in these patients. Sildenafil administration was also associated with increased vasodilator response during cold pressor testing, and an $18 \pm 6\%$ decrease in pulmonary artery pressure when compared with baseline values. In a separate cohort of 24 subjects with coronary artery disease and exercise-induced ischemia, administration of sildenafil 100 mg significantly increased flow-mediated dilation (FMD) in the brachial artery when compared with baseline, and slightly but not significantly increased exercise duration to ischemic threshold when compared with placebo.

Congestive heart failure

In a canine model of heart failure induced by rapid ventricular pacing, acute administration of sildenafil did not augment coronary vasodilator responses to either acetylcholine or nitroprusside [68]. PDE5 expression in ventricular myocardium homogenate was decreased in this experimental model of heart failure when compared with normal controls. The clinical implications of these findings are uncertain as the changes in coronary flow patterns during rapid ventricular pacing do not closely mimic human disease. Katz and colleagues studied the acute effects of sildenafil on flow-mediated dilation in the brachial artery in a double-blind, placebo-controlled prospective study of 48 subjects with chronic heart failure [69]. Flow-mediated dilation was measured with high-resolution ultrasound imaging after 1, 3, and 5 min of arterial occlusion before and 1 hour after administration of placebo or sildenafil at doses of 12.5 mg, 25 mg, or 50 mg. Sildenafil at doses of 25 and 50 mg significantly increased flow-mediated dilation when compared with placebo. Sildenafil was well-tolerated with no significant changes in blood pressure or heart rate when compared with placebo.

Diabetes mellitus

Fonseca and colleagues studied the effects of acute and chronic sildenafil administration on flow-mediated dilation in the brachial artery in 16 subjects with Type II diabetes mellitus in a double-blind, placebo-controlled crossover trial [70]. Sildenafil 25 mg significantly increased flow-mediated dilation after the first dose and 24 hours after the last dose of two weeks of daily doses of 25 mg when compared with placebo. No adverse events were reported. Heart rate and blood pressure effects were not reported in this study. In 24 diabetic subjects with coronary artery disease, sildenafil 50 mg was associated with a significant mean decrease in systolic blood pressure of 10 mmHg but no significant change in coronary flow reserve [71].

Pulmonary hypertension

Sildenafil has been shown to decrease pulmonary artery pressures, reduce right ventricular hypertrophy and ameliorate structural changes in experimental models of pulmonary hypertension induced by chronic hypoxia and administration of monocrotaline [72–74]. In mice with chronic hypoxia, the salutatory effect of sildenafil were partially maintained in knockout mouse strains without functioning endothelial nitric oxide synthase or functioning natriuretic peptide receptors, suggesting that the vasoactive and trophic effects of sildenafil may be related to cGMP signaling stimulated by both nitric oxide and natriuretic peptides [75, 76]. Sildenafil at doses ranging from 50–100 mg has been shown to attenuate hypoxia-induced pulmonary vasoconstriction in normal subjects, and acutely decrease pulmonary vascular resistance in patients with primary and secondary pulmonary hypertension [75, 77–79]. In a small pilot study of five patients with primary pulmonary hypertension, open-label administration of sildenafil 50 mg every eight hours for three months was associated with significant reductions in pulmonary vascular resistance, improved submaximal exercise capacity, and reduction in right ventricular mass when compared with baseline [80].

Cardiovascular safety of sildenafil

Cardiovascular side effects of sildenafil may be related to its vasorelaxant effects in vascular smooth muscle with consequent changes in coronary blood flow and systemic blood pressure. In experimental models and clinical studies of patients with coronary artery disease and diabetes mellitus, sildenafil induces mild vasodilation effects in the coronary circulation and is not associated with a reduction in coronary flow reserve [71, 81]. These findings in the catheterization laboratory have been confirmed in placebo-controlled exercise trials in patients with coronary artery disease and exercise induced ischemia. In a randomized crossover study of 105 subjects, sildenafil at doses of 50–100 mg significantly reduced resting blood pressure by a mean of 4 mmHg when compared with placebo but did not change exercise heart rate, exercise blood pressure, or exercise tolerance when compared with placebo [82]. In a double-blind, placebo-controlled parallel group study of 144 subjects, sildenafil 100 mg was associated with increased time to onset of angina, increased time to limiting angina, and increased exercise time when compared with placebo [83]. In 24 subjects with CHF with mixed underlying etiology, sildenafil 50 mg acutely improved exercise duration and peak oxygen uptake when compared with placebo. Oral administration of vardenafil also did not impair the ability to exercise at levels equivalent or greater than that attained during sexual activity when compared with placebo [84]. Interestingly, Ockaili et al. reported a pronounced infarct size-reducing effect of 0.7 mg/kg sildenafil in an *in vivo* rabbit model of coronary occlusion [85, 86]. As this effect could be blocked by 5-hydroxydecanoate, an inhibitor of mitochondrial ATP-sensitive potassium channels, it was suggested that cardioprotection by sildenafil may be attributed to activation of these potassium channels. Reffelman and Kloner also investigated the effects of sildenafil in a rabbit model of acute myocardial infarction (MI) and observed vasodilatory effects but no change in myocardial infarct size [87].

Histological studies indicate that the distribution of PDE5 in human myocardium is limited to vascular smooth muscle [88], but pharmacological studies have yielded conflicting findings. Wallis and colleagues demonstrated that sildenafil selectively increased cGMP levels, with no effect on cAMP levels or on contractile function in isolated human ventricular myocardium [57]. Immunohistochemical staining revealed that PDE5 expression was limited to vascular smooth muscle and was not present in myocytes. Stief et al. demonstrated increasing cAMP and cGMP concentrations in response to sildenafil $0.01-1$ μM in tissue obtained from isolated tissue from human atrial myocardium obtained at the time of open-heart surgery [89]. Senzaki and colleague reported that acute PDE5 inhibition with EMD82639 acutely attenuated the contractile response to dobutamine in normal dogs but not in dogs with pacing-induced heart failure [90]. These investigators demonstrated positive immunohistochemical staining for PDE5 in both vascular smooth muscle and myocardial cells in these canine hearts, with decreased PDE5 expression and enzyme activity in heart failure animals when compared with normal controls. Corbin and colleagues reported no direct effects of sildenafil 1 μM on contractile function in isolated myocardium from canine and human atrium [91]. PDE5 expression in homogenized atrial tissue was very low, consistent with the known distribution of PDE5 primarily within vascular smooth muscle. Species differences in PDE5 distribution and use of different experimental models and conditions may partly account for these divergent findings. There is no clear evidence to support direct myocardial effects of PDE5 inhibition in clinical circumstances [56]. In hemodynamic studies of patients with ischemic heart disease, chronic heart failure, and pulmonary hypertension, the hemodynamic profile of sildenafil resembles that of pure nitrosovasodilators [56, 77, 92, 93].

Clinical experience with sildenafil in patients with erectile dysfunction has demonstrated an excellent safety profile. In pooled data from 120 clinical trials with sildenafil, the rate of MI or cardiovascular death was 0.91 (95% confidence interval (CI) 0.52–1.48) events per 100 patient-years in subjects treated with sildenafil compared with 0.84 (95% CI 0.39–1.60) events per 100 patient-years in placebo-recipient patients. The relative risk of myocardial infarction or cardiovascular death was 1.08 (95% CI 0.45–2.77, p = 0.88) [94]. Post-marketing prescription event monitoring has confirmed the low rate of cardiovascular events in patients with erectile dysfunction and treated with sildenafil [95], and a recent report from the US Food and Drug Administration (FDA) did not find a higher incidence of cardiac events than what would have been expected for the patient population [96]. The safety and efficacy of sildenafil has also been demonstrated in controlled trials of smaller high risk populations of male patients with erectile dysfunction and hypertension on multiple anti-hypertensive agents, stable symptomatic coronary artery disease, and CHF [97–100]. Analysis of adverse events in controlled clinical trials of tadalafil also suggested no excess risk of cardiovascular events, although the clinical trial database and post-marketing safety database for these newer agents are much smaller than that available for sildenafil [101].

Conclusions

Endothelial dysfunction is highly prevalent in patients with established atherosclerotic cardiovascular disease, populations with traditional and non-traditional risk factors for cardiovascular disease, and patients with CHF. Endothelial dysfunction is linked to progression of atherosclerosis and heart failure and is associated with poor clinical outcomes. Preliminary studies in patients with coronary artery disease, CHF, diabetes mellitus and pulmonary hypertension indicate that PDE5 inhibition with sildenafil is well-tolerated and associated with acute improvements in endothelial function. Based on these encouraging findings, additional studies are warranted to further characterize the safety and efficacy of type 5 phosphodiesterase inhibition as part of a broad-based strategy to enhance vascular NO–cGMP signaling in patients with cardiovascular disease.

References

1 Vane JR, Anggard EE, Botting RM (1990) Regulatory functions of the vascular endothelium. *N Engl J Med* 323: 27–36
2 Rubanyi GM (1991) Endothelium-derived relaxing and contracting factors. *J Cell Biochem* 46: 27–36
3 Palmer RMJ, Ferrige AG, Moncada S (1987) Nitric oxide release accounts for the biological activity of endothelium-derived relaxing factor. *Nature* 327: 524–526
4 Moncada S, Palmer RMJ, Higgs EA (1991) Nitric oxide: physiology, pathophysiology, and pharmacology. *Pharmacol Rev* 43: 109–142
5 Lincoln TM (1989) Cyclic GMP and mechanisms of vasodilation. *Pharmacol Ther* 41: 479–502
6 Lucas KA, Pitari GM, Kazerounian S, Ruiz-Stewart I, Park J, Schulz S, Chepenik KP, Waldman SA (2000) Guanylyl cyclases and signaling by cyclic GMP. *Pharmacol Rev* 52: 375–414
7 Beavo JA (1995) Cyclic nucleotide phosphodiesterases: functional implications of multiple isoforms. *Physiol Rev* 75: 725–748
8 Corbin JD, Francis SH (1999) Cyclic GMP phosphodiesterase-5: target of sildenafil. *J Biol Chem* 274: 13729–13732

9. Corbin JD, Francis SH (2002) Pharmacology of phosphodiesterase-5 inhibitors. *Int J Clin Pract* 56: 453–459
10. Turko IV, Ballard SA, Francis SH, Corbin JD (1999) Inhibition of cyclic GMP-binding cyclic GMP-specific phosphodiesterase (Type 5) by sildenafil and related compounds. *Mol Pharmacol* 56: 124–130
11. Rybalkin SD, Rybalkina IG, Feil R, Hofmann F, Beavo JA (2002) Regulation of cGMP-specific phosphodiesterase (PDE5) phosphorylation in smooth muscle cells. *J Biol Chem* 277: 3310–3317
12. Rybalkin SD, Rybalkina IG, Shimizu-Albergine M, Tang XB, Beavo JA (2003) PDE5 is converted to an activated state upon cGMP binding to the GAF A domain. *Embo J* 22: 469–478
13. Mullershausen F, Russwurm M, Thompson WJ, Liu L, Koesling D, Friebe A (2001) Rapid nitric oxide-induced desensitization of the cGMP response is caused by increased activity of phosphodiesterase type 5 paralleled by phosphorylation of the enzyme. *J Cell Biol* 155: 271–278
14. Francis SH, Bessay EP, Kotera J, Grimes KA, Liu L, Thompson WJ, Corbin JD (2002) Phosphorylation of isolated human phosphodiesterase-5 regulatory domain induces an apparent conformational change and increases cGMP binding affinity. *J Biol Chem* 277: 47581–47587
15. Corbin JD, Francis SH, Webb DJ (2002) Phosphodiesterase type 5 as a pharmacologic target in erectile dysfunction. *Urology* 60: 4–11
16. Furchgott RF, Zawadzki JV (1980) The obligatory role of endothelial cells in the relaxation of arterial smooth muscle by acetylcholine. *Nature* 288: 374–376
17. Loscalzo J, Vita JA (1994) Ischemia, hyperemia, exercise, and nitric oxide. Complex physiology and complex molecular adaptations [editorial; comment]. *Circulation* 90: 2556–2559
18. Stamler JS, Singel DJ, Loscalzo J (1992) Biochemistry of nitric oxide and its redox-activated forms. *Science* 258: 1898–1902
19. Fleming I, Busse R (1999) Signal transduction of eNOS activation. *Cardiovasc Res* 43: 532–541
20. Fulton D, Gratton JP, Sessa WC (2001) Post-translational control of endothelial nitric oxide synthase: why isn't calcium/calmodulin enough? *J Pharmacol Exp Ther* 299: 818–824
21. Ludmer PL, Selwyn AP, Shook TL, Wayne RR, Mudge GH, Alexander RW, Ganz P (1986) Paradoxical vasoconstriction induced by acetylcholine in atherosclerotic coronary arteries. *N Engl J Med* 315: 1046–1051
22. Libby P (2001) Current concepts of the pathogenesis of the acute coronary syndromes. *Circulation* 104: 365–372
23. Pepine CJ (1998) Clinical implications of endothelial dysfunction. *Clin Cardiol* 21: 795–799
24. Zheng H, Dimayuga C, Hudaihed A, Katz SD (2002) Effect of dexrazoxane on homocysteine-induced endothelial dysfunction in normal subjects. *Arterioscler Thromb Vasc Biol* 22: E15–18
25. Vogel RA, Corretti MC, Plotnick GD (1997) Effect of a single high-fat meal on endothelial function in healthy subjects. *Am J Cardiol* 79: 350–354
26. Harrison DG (1997) Cellular and molecular mechanisms of endothelial cell dysfunction. *J Clin Invest* 100: 2153–2157
27. Katz SD, Khan T, Zeballos GA, Mathew L, Potharlanka P, Knecht M, Whelan J (1999) Decreased activity of the L-arginine–nitric oxide metabolic pathway in patients with congestive heart failure. *Circulation* 99: 2118–2123
28. Kojda G, Harrison D (1999) Interactions between NO and reactive oxygen species: pathophysiological importance in atherosclerosis, hypertension, diabetes and heart failure. *Cardiovasc Res* 43: 562–571
29. Wolin MS (2000) Interactions of oxidants with vascular signaling systems. *Arterioscler Thromb Vasc Biol* 20: 1430–1442
30. Hornig B, Arakawa N, Kohler C, Drexler H (1998) Vitamin C improves endothelial function of conduit arteries in patients with chronic heart failure. *Circulation* 97: 363–368
31. Heitzer T, Brockhoff C, Mayer B, Warnholtz A, Mollnau H, Henne S, Meinertz T, Munzel T (2000) Tetrahydrobiopterin improves endothelium-dependent vasodilation in chronic smokers: evidence for a dysfunctional nitric oxide synthase. *Circ Res* 86: E36–41
32. Ross R (1999) Atherosclerosis – an inflammatory disease. *N Engl J Med* 340: 115–126
33. Mancini GB (2004) Vascular structure *versus* function: is endothelial dysfunction of independent prognostic importance or not? *J Am Coll Cardiol* 43: 624–628
34. Katz SD, Zheng H (2002) Peripheral limitations of maximal aerobic capacity in patients with chronic heart failure. *J Nucl Cardiol* 9: 215–225
35. Zelis R, Longhurst J, Capone RJ, Lee G (1973) Peripheral circulatory control mechanisms in congestive heart failure. *Am J Cardiol* 32: 481–490

36 LeJemtel TH, Maskin CS, Lucido D, Chadwick BJ (1986) Failure to augment maximal blood flow in response to one-leg *versus* two-leg exercise in patients with congestive heart failure. *Circulation* 74: 245–251
37 Jondeau G, Katz SD, Zohman LR, Goldberger M, McCarthy M, Boudarias JP, LeJemtel TH (1992) Active skeletal muscle mass and cardiopulmonary reserve: Failure to attain peak aerobic capacity during maximal exercise in patients with congestive heart failure. *Circulation* 86: 1351–1356
38 Zelis R, Mason DT, Braunwald E (1968) A comparison of the effects of vasodilator stimuli on peripheral resistance vessels in normal subjects and in patients with congestive heart failure. *J Clin Invest* 47: 960–970
39 Zelis R, Longhurst J, Capone RJ, Mason DT (1974) A comparison of regional blood flow and oxygen utilization during dynamic forearm exercise in normal subjects and patients with congestive heart failure. *Circulation* 50: 137–143
40 Katz SD (1995) The role of endothelium-derived vasoactive substances in the pathophysiology of exercise intolerance in patients with congestive heart failure. *Prog Cardiovasc Dis* 28: 23–50
41 Hirai T, Zelis R, Musch TI (1995) Effects of nitric oxide synthase inhibition on the muscle blood flow response to exercise in rats with heart failure. *Cardiovasc Res* 30: 469–476
42 Katz SD, Krum H, Khan T, Knecht M (1996) Exercise-induced vasodilation in forearm circulation of normal subjects and patients with congestive heart failure: role of endothelium-derived nitric oxide. *J Am Coll Cardiol* 28: 585–590
43 Nakamura M, Ishikawa M, Funakoshi T, Hashimoto K, Chiba M, Hiramori K (1994) Attenuated endothelium-dependent peripheral vasodilation and clinical characteristics in patients with chronic heart failure. *Am Heart J* 128: 1164–1169
44 Nakamura M, Yoshida H, Arakawa N, Mizunuma Y, Makita S, Hiramori K (1996) Endothelium-dependent vasodilatation is not selectively impaired in patients with chronic heart failure secondary to valvular heart disease and congenital heart disease. *Eur Heart J* 17: 1875–1881
45 Carville C, Adnot S, Sediame S, Benacerraf S, Castaigne A, Calvo F, de Cremou P, Dubois-Rande JL (1998) Relation between impairment in nitric oxide pathway and clinical status in patients with congestive heart failure. *J Cardiovasc Pharmacol* 32: 562–570
46 Bauersachs J, Bouloumie A, Fraccarollo D, Hu K, Busse R, Ertl G (1999) Endothelial dysfunction in chronic myocardial infarction despite increased vascular endothelial nitric oxide synthase and soluble guanylate cyclase expression: role of enhanced vascular superoxide production. *Circulation* 100: 292–298
47 Maguire SM, Nugent AG, McGurk C, Johnston GD, Nicholls DP (1998) Abnormal vascular responses in human chronic cardiac failure are both endothelium dependent and endothelium independent. *Heart* 80: 141–145
48 Nakamura M, Yoshida H, Arakawa N, Mizunuma Y, Makita S, Hiramori K (1996) Endothelium-dependent vasodilatation is not selectively impaired in patients with chronic heart failure secondary to valvular heart disease and congenital heart disease. *Eur Heart J* 17: 1875–1881
49 Nakamura M, Arakawa N, Yoshida H, Makita S, Niinuma H, Hiramori K (1998) Vasodilatory effects of B-type natriuretic peptide are impaired in patients with chronic heart failure. *Am Heart J* 135: 414–420
50 Creager MA, Quigg RJ, Ren CJ, Roddy MA, Colucci WS (1991) Limb vascular responsiveness to beta-adrenergic receptor stimulation in patients with congestive heart failure. *Circulation* 83: 1873–1879
51 Katz SD, Schwarz M, Yuen J, LeJemtel TH (1993) Impaired acetylcholine-mediated vasodilation in patients with congestive heart failure. Role of endothelium-derived vasodilating and vasoconstricting factors. *Circulation* 88: 55–61
52 Supaporn T, Sandberg SM, Borgeson DD, Heublein DM, Luchner A, Wei CM, Dousa TP, Burnett JC, Jr. (1996) Blunted cGMP response to agonists and enhanced glomerular cyclic 3',5'-nucleotide phosphodiesterase activities in experimental congestive heart failure. *Kidney Int* 50: 1718–1725
53 Haneda M, Kikkawa R, Maeda S, Togawa M, Koya D, Horide N, Kajiwara N, Shigeta Y (1991) Dual mechanism of angiotensin II inhibits ANP-induced mesangial cGMP accumulation. *Kidney Int* 40: 188–194
54 Maurice DH, Palmer D, Tilley DG, Dunkerley HA, Netherton SJ, Raymond DR, Elbatarny HS, Jimmo SL (2003) Cyclic nucleotide phosphodiesterase activity, expression, and targeting in cells of the cardiovascular system. *Mol Pharmacol* 64: 533–546

55 Francis SH, Lincoln TM, Corbin JD (1980) Characterization of a novel cGMP binding protein from rat lung. *J Biol Chem* 255: 620–626
56 Reffelmann T, Kloner RA (2003) Therapeutic potential of phosphodiesterase 5 inhibition for cardiovascular disease. *Circulation* 108: 239–244
57 Wallis RM, Corbin JD, Francis SH, Ellis P (1999) Tissue distribution of phosphodiesterase families and the effects of sildenafil on tissue cyclic nucleotides, platelet function, and the contractile responses of trabeculae carneae and aortic rings *in vitro*. *Am J Cardiol* 83: 3C–12C
58 Juilfs DM, Soderling S, Burns F, Beavo JA (1999) Cyclic GMP as substrate and regulator of cyclic nucleotide phosphodiesterases (PDEs). *Rev Physiol Biochem Pharmacol* 135: 67–104
59 Ballard SA, Gingell CJ, Tang K, Turner LA, Price ME, Naylor AM (1998) Effects of sildenafil on the relaxation of human corpus cavernosum tissue *in vitro* and on the activities of cyclic nucleotide phosphodiesterase isozymes. *J Urol* 159: 2164–2171
60 Goldstein I, Lue TF, Padma-Nathan H, Rosen RC, Steers WD, Wicker PA (1998) Oral sildenafil in the treatment of erectile dysfunction. Sildenafil Study Group [published erratum appears in N Engl J Med 1998 Jul 2: 339(1):59]. *N Engl J Med* 338: 1397–1404
61 Hellstrom WJ, Gittelman M, Karlin G, Segerson T, Thibonnier M, Taylor T, Padma-Nathan H (2003) Sustained efficacy and tolerability of vardenafil, a highly potent selective phosphodiesterase type 5 inhibitor, in men with erectile dysfunction: results of a randomized, double-blind, 26-week placebo-controlled pivotal trial. *Urology* 61: 8–14
62 Gresser U, Gleiter CH (2002) Erectile dysfunction: comparison of efficacy and side effects of the PDE-5 inhibitors sildenafil, vardenafil and tadalafil – review of the literature. *Eur J Med Res* 7: 435–446
63 Kim NN, Huang YH, Goldstein I, Bischoff E, Trais AM (2001) Inhibition of cyclic GMP hydrolysis in human corpus cavernosum smooth muscle cells by vardenafil, a novel, selective phosphodiesterase type 5 inhibitor. *Life Sci* 69: 2249–2256
64 Porst H, Padma-Nathan H, Giuliano F, Anglin G, Varanese L, Rosen R (2003) Efficacy of tadalafil for the treatment of erectile dysfunction at 24 and 36 hours after dosing: a randomized controlled trial. *Urology* 62: 121–125; discussion 125–126
65 Brock GB, McMahon CG, Chen KK, Costigan T, Shen W, Watkins V, Anglin G, Whitaker S (2002) Efficacy and safety of tadalafil for the treatment of erectile dysfunction: results of integrated analyses. *J Urol* 168: 1332–1336
66 Traverse JH, Chen YJ, Du R, Bache RJ (2000) Cyclic nucleotide phosphodiesterase type 5 activity limits blood flow to hypoperfused myocardium during exercise. *Circulation* 102: 2997–3002
67 Halcox JP, Nour KR, Zalos G, Mincemoyer RA, Waclawiw M, Rivera CE, Willie G, Ellahham S, Quyyumi AA (2002) The effect of sildenafil on human vascular function, platelet activation, and myocardial ischemia. *J Am Coll Cardiol* 40: 1232–1240
68 Chen Y, Traverse JH, Hou M, Li Y, Du R, Bache RJ (2003) Effect of PDE5 inhibition on coronary hemodynamics in pacing-induced heart failure. *Am J Physiol Heart Circ Physiol* 284: H1513–1520
69 Katz SD, Balidemaj K, Homma S, Wu H, Wang J, Maybaum S (2000) Acute type 5 phosphodiesterase inhibition with sildenafil enhances flow-mediated vasodilation in patients with chronic heart failure. *J Am Coll Cardiol* 36: 845–851
70 Fonseca V (2003) Acute and prolonged effects of sildenafil on brachial artery flow-mediated dilatation in type 2 diabetes: response to khaodhiar and veves. *Diabetes Care* 26: 963
71 Dietz U, Tries HP, Merkle W, Jaursch-Hancke C, Lambertz H (2003) Coronary artery flow reserve in diabetics with erectile dysfunction using sildenafil. *Cardiovasc Diabetol* 2: 8
72 Schermuly RT, Kreisselmeier KP, Ghofrani HA, Yilmaz H, Butrous G, Ermert L, Ermert M, Weissmann N, Rose F, Guenther A et al. (2004) Chronic sildenafil treatment inhibits monocrotaline-induced pulmonary hypertension in rats. *Am J Respir Crit Care Med* 169: 39–45
73 Itoh T, Nagaya N, Fujii T, Iwase T, Nakanishi N, Hamada K, Kangawa K, Kimura H (2004) A combination of oral sildenafil and beraprost ameliorates pulmonary hypertension in rats. *Am J Respir Crit Care Med* 169: 34–38
74 Sebkhi A, Strange JW, Phillips SC, Wharton J, Wilkins MR (2003) Phosphodiesterase type 5 as a target for the treatment of hypoxia-induced pulmonary hypertension. *Circulation* 107: 3230–3235
75 Zhao L, Mason NA, Morrell NW, Kojonazarov B, Sadykov A, Maripov A, Mirrakhimov MM, Aldashev A, Wilkins MR (2001) Sildenafil inhibits hypoxia-induced pulmonary hypertension. *Circulation* 104: 424–428
76 Zhao L, Mason NA, Strange JW, Walker H, Wilkins MR (2003) Beneficial effects of phosphodi-

esterase 5 inhibition in pulmonary hypertension are influenced by natriuretic Peptide activity. *Circulation* 107: 234–237
77 Lepore JJ, Maroo A, Pereira NL, Ginns LC, Dec GW, Zapol WM, Bloch KD, Semigran MJ (2002) Effect of sildenafil on the acute pulmonary vasodilator response to inhaled nitric oxide in adults with primary pulmonary hypertension. *Am J Cardiol* 90: 677–680
78 Michelakis E, Tymchak W, Lien D, Webster L, Hashimoto K, Archer S (2002) Oral sildenafil is an effective and specific pulmonary vasodilator in patients with pulmonary arterial hypertension: comparison with inhaled nitric oxide. *Circulation* 105: 2398–2403
79 Leuchte HH, Schwaiblmair M, Baumgartner RA, Neurohr CF, Kolbe T, Behr J (2004) Hemodynamic response to sildenafil, nitric oxide, and iloprost in primary pulmonary hypertension. *Chest* 125: 580–586
80 Michelakis ED, Tymchak W, Noga M, Webster L, Wu XC, Lien D, Wang SH, Modry D, Archer SL (2003) Long-term treatment with oral sildenafil is safe and improves functional capacity and hemodynamics in patients with pulmonary arterial hypertension. *Circulation* 108: 2066–2069
81 Herrmann HC, Chang G, Klugherz BD, Mahoney PD (2000) Hemodynamic effects of sildenafil in men with severe coronary artery disease. *N Engl J Med* 342: 1622–1626
82 Arruda-Olson AM, Mahoney DW, Nehra A, Leckel M, Pellikka PA (2002) Cardiovascular effects of sildenafil during exercise in men with known or probable coronary artery disease: a randomized crossover trial. *JAMA* 287: 719–725
83 Fox KM, Thadani U, Ma PT, Nash SD, Keating Z, Czorniak MA, Gillies H, Keltai M (2003) Sildenafil citrate does not reduce exercise tolerance in men with erectile dysfunction and chronic stable angina. *Eur Heart J* 24: 2206–2212
84 Thadani U, Smith W, Nash S, Bittar N, Glasser S, Narayan P, Stein RA, Larkin S, Mazzu A, Tota R et al. (2002) The effect of vardenafil, a potent and highly selective phosphodiesterase-5 inhibitor for the treatment of erectile dysfunction, on the cardiovascular response to exercise in patients with coronary artery disease. *J Am Coll Cardiol* 40: 2006–2012
85 Kukreja RC, Ockaili R, Salloum F, Yin C, Hawkins J, Das A, Xi L (2004) Cardioprotection with phosphodiesterase-5 inhibition – a novel preconditioning strategy. *J Mol Cell Cardiol* 36: 165–173
86 Ockaili R, Salloum F, Hawkins J, Kukreja RC (2002) Sildenafil (Viagra) induces powerful cardioprotective effect via opening of mitochondrial K(ATP) channels in rabbits. *Am J Physiol Heart Circ Physiol* 283: H1263–1269
87 Reffelmann T, Kloner RA (2003) Effects of sildenafil on myocardial infarct size, microvascular function, and acute ischemic left ventricular dilation. *Cardiovasc Res* 59: 441–449
88 Katz SD, Parums DV (2001) Sympathetic activation by sildenafil. *Circulation* 104: E119–120
89 Stief CG, Uckert S, Becker AJ, Harringer W, Truss MC, Forssmann WG, Jonas U (2000) Effects of sildenafil on cAMP and cGMP levels in isolated human cavernous and cardiac tissue. *Urology* 55: 146–150
90 Senzaki H, Smith CJ, Juang GJ, Isoda T, Mayer SP, Ohler A, Paolocci N, Tomaselli GF, Hare JM, Kass DA (2001) Cardiac phosphodiesterase 5 (cGMP-specific) modulates beta-adrenergic signaling *in vivo* and is down-regulated in heart failure. *FASEB J* 15: 1718–1726
91 Corbin J, Rannels S, Neal D, Chang P, Grimes K, Beasley A, Francis S (2003) Sildenafil citrate does not affect cardiac contractility in human or dog heart. *Curr Med Res Opin* 19: 747–752
92 Jackson G, Benjamin N, Jackson N, Allen MJ (1999) Effects of sildenafil citrate on human hemodynamics. *Am J Cardiol* 83: 13C–20C
93 Manfroi WC, Caramori PR, Zago AJ, Melchior R, Zen V, Accordi M, Gutierres D, Noer C (2003) Hemodynamic effects of sildenafil in patients with stable ischemic heart disease. *Int J Cardiol* 90: 153–157
94 Mittleman MA, Glasser DB, Orazem J (2003) Clinical trials of sildenafil citrate (Viagra) demonstrate no increase in risk of myocardial infarction and cardiovascular death compared with placebo. *Int J Clin Pract* 57: 597–600
95 Shakir SA, Wilton LV, Boshier A, Layton D, Heeley E (2001) Cardiovascular events in users of sildenafil: results from first phase of prescription event monitoring in England. *BMJ* 322: 651–652
96 Wysowski DK, Farinas E, Swartz L (2002) Comparison of reported and expected deaths in sildenafil (Viagra) users. *Am J Cardiol* 89: 1331–1334
97 DeBusk RF, Pepine CJ, Glasser DB, Shpilsky A, DeRiesthal H, Sweeney M (2004) Efficacy and safety of sildenafil citrate in men with erectile dysfunction and stable coronary artery disease. *Am J Cardiol* 93: 147–153

98 Zusman RM, Prisant LM, Brown MJ (2000) Effect of sildenafil citrate on blood pressure and heart rate in men with erectile dysfunction taking concomitant antihypertensive medication. Sildenafil Study Group. *J Hypertens* 18: 1865–1869
99 Webster LJ, Michelakis ED, Davis T, Archer SL (2004) Use of Sildenafil for Safe Improvement of Erectile Function and Quality of Life in Men With New York Heart Association Classes II and III Congestive Heart Failure: A Prospective, Placebo-Controlled, Double-blind Crossover Trial. *Arch Intern Med* 164: 514–520
100 Bocchi EA, Guimaraes G, Mocelin A, Bacal F, Bellotti G, Ramires JF (2002) Sildenafil effects on exercise, neurohormonal activation, and erectile dysfunction in congestive heart failure: a double-blind, placebo-controlled, randomized study followed by a prospective treatment for erectile dysfunction. *Circulation* 106: 1097–1103
101 Kloner RA, Mitchell M, Emmick JT (2003) Cardiovascular effects of tadalafil. *Am J Cardiol* 92: 37 M–46 M

Sildenafil and cardiovascular events – drug interactions

Graham Jackson

Cardiothoracic Centre, St Thomas Hospital, Lambeth Palace Road, London SE1 7EH, UK

Introduction

The most common cause of erectile dysfunction (ED) is vascular endothelial dysfunction [1]. Cardiovascular diseases share the same risk factors as ED – smoking, hyperlipidaemia, diabetes and hypertension – and the coexistence of a cardiac problem may cause concern with regard to advising on the treatment of ED (Fig. 1). This concern has increased with the realisation that ED is a marker for previously undetected coronary artery disease and the casual use of sildenafil without proper assessment – so-called 'internet sex' – precipitating acute cardiac events [2, 3].

In the era of evidence-based medicine, in order for all patients with coronary disease to maximise risk reduction they should be taking aspirin or clopidogrel, a statin, a beta blocker and an angiotensin-converting-enzyme inhibitor or angiotensin II antagonist [4]. They may be on additional medication

Risk Factors

Coronary Heart Disease		Erectile Dysfunction
Smoking		Smoking
Blood Pressure		Blood Pressure
Cholesterol ← E.D. →		Cholesterol
Diabetes		Diabetes

Endothelial Dysfunction - the common denominator

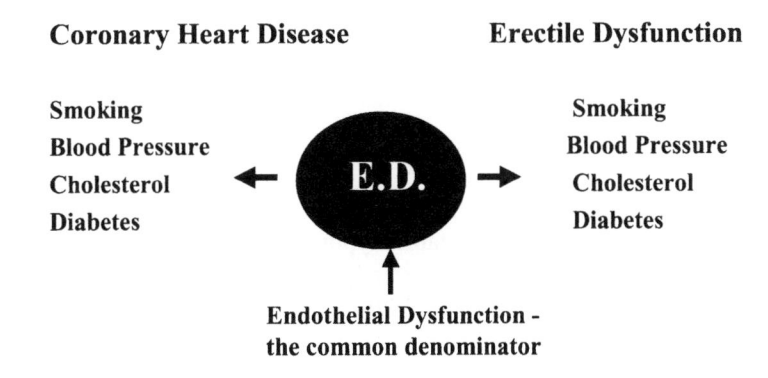

Figure 1. Erectile dysfunction and cardiovascular disease share the same risk factors, and endothelial dysfunction is the common denominator. From [1], with permission.

depending on their underlying problems or co-morbidities. The possibility of drug interactions is always a consideration when prescribing in general but will need additional thought with the intermittent use of sildenafil or the other phosphodiesterase type 5 (PDE5) inhibitors tadalafil and vardenafil. The casual user of sildenafil may be unaware of the possibility of drug interactions and present a clinical challenge if its use is not volunteered.

In this Chapter I will review the information we have on the presence or absence of significant PDE5 inhibitor drug interactions.

PDE5 Inhibitors – mode of action

This has already been reviewed in this volume. However it is important to realise that these drugs are arterial and venous vasodilators that act by blocking the degradation of cyclic guanosine monophosphates (cGMP) – a process present through the vascular tree, not just in the penis [5]. As nitric oxide is the key driver of this relationship an additive effect with a nitric oxide donor may significantly dilate the vascular circuit leading to clinically important hypotension (Fig. 2). This potential interaction will be present for all PDE5 inhibitors but will vary in duration depending on each drug's half-life (Tab. 1). Given the equipotency of the drugs in ED terms, from the cardiovascular point of view their adverse event potential will be determined by the combination of dosage, duration of action and co-prescribing.

Table 1. PDE5 inhibitors: essential differences

	Half-life (hrs)	Maximal effect (mins)	Duration (hrs)
Sildenafil	4–6	30–60	4–6+
Vardenafil	4–6	30–60	4–6+
Tadalafil	17.5	60–120	36–48

Nitrates

Organic nitrates represent a specific contraindication to the use of all PDE5 inhibitors because of an unpredictable hypotensive interaction with systolic blood pressures falling abruptly by over 30 mmHg, or to below 85 mmHg. Sublingual nitrates are not advised 12 hours before or after sildenafil or vardenafil use, and 48 hours after tadalafil use [6, 7]. Nitrogen oxide (NO) donors act by catalysing the formation of cGMP which may be increased significantly by a PDE5 inhibitor leading to profound vasodilatation (Fig. 2b) as cGMP build up is facilitated throughout the vascular tree. Sildenafil alone causes only

a

Nitrates → NO

Naturiuretic peptides

NO → Guanylate cyclase ← Naturiuretic peptides

GTP →(Activation)→ cGMP →(Degradation)→ GMP

PDE5 ↑

cGMP ┈┈▶ Smooth Muscle Relaxation

b

Nitrates → NO

Naturiuretic peptides

NO → Guanylate cyclase ← Naturiuretic peptides

GTP →(Activation)→ cGMP →(Degradation)→ GMP

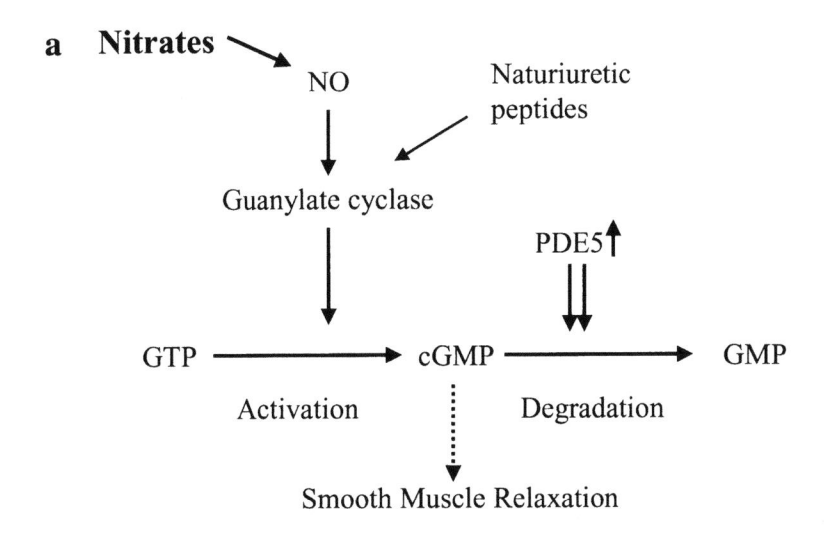

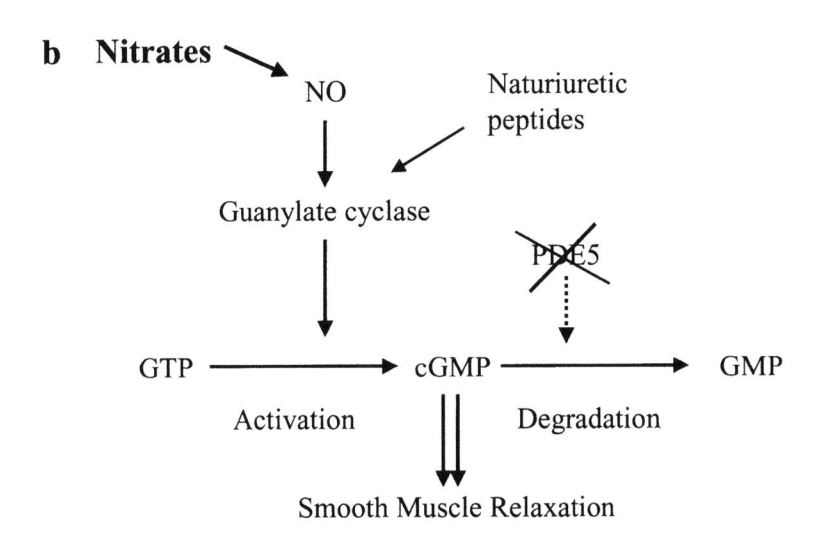

~~PDE5~~

cGMP ⇒ Smooth Muscle Relaxation

Figure 2. (a) The normal nitric oxide pathway. (b) The potential impact of nitrates in the presence of a PDE5 inhibitor.

a mild reduction in blood pressure (–8/–5.5 mmHg) but following sublingual glyceryl trinitrate falls of between –26 and –56 mmHg in systolic pressure have been recorded [8].

Oral isosorbide mononitrate similarly has the potential to decrease blood pressure in the presence of sildenafil with one study recording falls of up to 29/15 mmHg compared to placebo [9]. Oral nitrates should be discontinued five half-lives before sildenafil use, which in effect is five days with the popular long-acting nitrate formulations.

Vardenafil has a similar half-life to sildenafil at about 4 h and can be considered similar in drug interaction terms regarding nitrates [10].

Tadalafil with its half-life of 17.5 h, while having similar hypotensive effects to sildenafil in the absence of nitrates, has a potentially prolonged interaction to sublingual nitrates, which resolves by 48 h [7]. Not all patients become hypotensive but there are unpredictable "outliers" who in comparison to placebo sustain falls in systolic pressure greater than 30 mmHg or to a systolic pressure less than 85 mmHg. Once more, unpredictable outliers influence the recommendation with oral nitrates, which should not be co-prescribed in the five days time window.

The use of intravenous nitrates in the emergency setting is a subject of debate with no clear evidence based guidelines. In the out-of-hospital environment, as with sublingual nitrates, they should not be used in an acute setting following the use of a PDE5 inhibitor (patients must be told to volunteer PDE5 inhibitor use, paramedics must ask) but in the emergency room or coronary unit with close monitoring intravenous nitrates may be used as they can be immediately dose adjusted or terminated.

Amyl nitrite or amyl nitrate ("poppers") are used to enhance sexual activity, particularly in the homosexual community during anal intercourse. A potentially similar PDE5 inhibitor nitrate interaction can occur. As "poppers" may be used in a confined environment (e.g., a sauna) by a person other than the PDE5 inhibitor user, the effect of spreading throughout the atmosphere can potentially affect the PDE5 inhibitor user. It is imperative that patient and partner(s) are advised appropriately so that inadvertent potentially dangerous exposure can be avoided.

Stopping nitrates

Oral nitrates are relatively weak anti-anginal drugs with no prognostic value and can be discontinued or substituted in most stable patients. In one series of 42 patients, 40 safely discontinued oral isosorbide mononitrate and were reviewed 1–2 weeks later. 36 were then effectively treated with a PDE5 inhibitor [11]. There were no adverse events or decrease in exercise ability. In five patients, amlodipine was successfully substituted.

A similar study found most people carrying sublingual nitrates rarely used them and with appropriate counselling it was possible to prescribe a PDE5 inhibitor [12]. Nitrate use should therefore not be an outright contraindication to PDE5 inhibition but a focus for re-evaluation of the cardiac patient.

Key points

- Nitrates and PDE5 inhibitors may unpredictably interact causing clinically important hypotension

- Sublingual nitrates are not recommended 12 h before a PDE5 inhibitor or within 12 h of sildenafil or vardenafil use and 48 h of tadalafil use
- Long acting (daily dosing) oral nitrates should not be used five days before PDE5 inhibitor use. Post PDE5 inhibitors can be started as for sublingual nitrates
- Oral nitrates can often be substituted to allow PDE5 inhibitor use

Nicorandil and nebivolol

These drugs act on the nitric oxide pathway. They have not been evaluated with PDE5 inhibitors. On theoretical grounds caution should be exercised and in the absence of safety data and the presence of alternative therapies they are currently not recommended for use with a PDE5 inhibitor [6].

Alpha blockers

The possibility of postural hypotension with co-prescribing doxazosin and sildenafil has led to a time separation recommendation of 4 h and for tadalafil 8 h [13]. Vardenafil is not recommended with any alpha blocker whereas tadalafil can be used with tamsulosin 0.4 mg.

Though the mean difference with tadalafil was less than 10 mmHg the number of outliers with an systolic blood pressure (SBP) <85 mmHg was greater after doxazosin plus tadalafil (28%) *versus* doxazosin plus placebo (6%). No subject taking tamsulosin had a decrease in SBP to less than 85 mmHg.

As doxazosin is used to treat benign prostatic hypertrophy, a population of patients vulnerable to ED, this interaction needs to be considered when PDE5 inhibitors are co-prescribed. Whether this occurs with the extended release formulation of doxazosin is unknown.

Safety with other cardiovascular drugs

Most information is available with sildenafil, but drug interaction studies with multiple antihypertensive therapy other than alpha blockers indicates safety with all three PDE5 inhibitors [14]. There is no evidence of a significant interaction with beta blockers, calcium antagonists, diuretics, angiotensin converting enzyme inhibitors or angiotensin II antagonists. Small differences in standing blood pressure of up to −7 mmHg systolic and −4 mmHg diastolic have been observed in controlled hypertensives on therapy and these were not clinically significant. The potential to lower blood pressure in uncontrolled hypertensives may open a therapeutic window for longer-acting PDE5 inhibitors such as tadalafil [15].

Drugs influencing excretion

The CYP3A4 inhibitor erythromycin increases the sildenafil area under the curve (AUC) by 182% but with intermittent sildenafil use this has no documented clinical significance [16]. Ritonavir and saquinavir, which inhibit HIV protease, do significantly increase sildenafil's peak levels and duration of action, and specialised advice is recommended when co-prescribing (one 25 mg dose per 48 h). Similar interactions are reported with vardenafil (erythromycin AUC four-fold increase, ketoconazolole 10-fold and indinavir 16-fold with maximum plasma concentrations (Cmax) increased 3.1, 4.1 and seven-fold, respectively). Tadalafil has no significant interactions at 10 mg other than the protease inhibitors but, given the similar metabolic pathways, similar cautions are needed at both 10 and 20 mg as sildenafil and vardenafil [14].

In clinical practice, with on demand use only the protease inhibitors merit specific advice. However, if regular daily dosing becomes routine practice, as I suspect it will with tadalafil, more detailed interaction studies will be needed.

PDE5 inhibitors and the future

With our better understanding of the role of the endothelium in erectile dysfunction and its fundamental importance to vascular biology, has come the recognition that drugs used to treat ED may have important therapeutic implications throughout the vasculature [17]. PDE5 inhibitors are now being seen as potentially important cardiovascular drugs for use in pulmonary hypertension, systemic hypertension, cardiac failure and Raynaud's phenomenon. The upregulation of PDE5 by the vasoconstrictor angiotensin II and its reversal by a PDE5 inhibitor could result in important therapeutic vasodilatory properties and the evidence of improved endothelial function both therapeutic and prognostic potential [17]. Currently PDE5 inhibitors are used on demand and the safety data is reassuring. However, if they become drugs to be used daily, long-term evaluation of safety and benefit, both symptomatic and prognostic, will become necessary.

References

1 Solomon H, Man JW, Jackson G (2003) Erectile dysfunction and the cardiovascular patient: endothelial dysfunction is the common denominator. *Heart* 89: 251–253
2 Kirby M, Jackson G, Betteridge J, Friedli K (2001) Is erectile dysfunction a marker for cardiovascular disease? *Int J Clin Pract* 55: 614–618
3 Solomon H, Man J, Jackson G (2002) Viagra on the Internet – Unsafe Sexual Practice. *Int J Clin Pract* 56: 403–404
4 Yusuf S (2002) Two decades of progress in preventing vascular disease. *Lancet* 360: 2–3
5 Gillies HC, Roblin D, Jackson G (2002) Coronary and systemic haemodynamic effects of sildenafil citrate: from basic science to clinical studies in patients with cardiovascular disease. *Int J Cardiol* 86: 131–141

6 Jackson G, Betteridge J, Dean J, Eardley I, Hall R, Holdright D, Holmes S, Kirby M, Riley A, Sever P (2002) A systematic approach to erectile dysfunction in the cardiovascular patient: a consensus statement – update 2002. *Int J Clin Pract* 56: 663–671
7 Kloner RA, Hutter AM, Emmick JT, Mitchell MI, Denne J, Jackson G (2003) Time course of interaction between tadalafil and nitrates. *J Am Coll Cardiol* 42: 1855–1860
8 Webb DJ, Freestone S, Allen MJ, Muirhead GJ (1999) Sildenafil citrate and blood pressure lowering drugs: results of drug interaction studies with an organic nitrate and calcium antagonist. *Am J Cardiol* 83: 21C–28C
9 Webb DJ, Muirhead GJ, Wulff M, Sutton JA, Levi R, Dinsmore WW (2000) Sildenafil citrate potentiates the hypotensive effects of nitric oxide donor drugs in male patients with stable angina. *J Am Coll Cardiol* 36: 25–31
10 Stief C, Porst H, Saenz De Tejada I, Ulbrich E, Beneke M; Vardenafil Study Group (2004) Sustained efficacy and tolerability with vardenafil over 2 years of treatment in men with erectile dysfunction. *Int J Clin Pract* 58: 230–239
11 Jackson G, Cooper A, McGing E, Martin E (2003) Successful withdrawal of oral long-acting nitrates to facilitate phosphodiesterase type 5 inhibitor use in cardiac patients with erectile dysfunction. *Int J Imp Res* 15 (Suppl 6): 520 (Abst)
12 Martin E, Cooper A, McGing E, Jackson G (2003) Carrying short acting nitrates is not a contraindication to PDE5 inhibition. *Int J Imp Res* 15 (Suppl 6): 520 (Abst)
13 Jackson G, Bedding A, Emmich J (2003) Interaction between tadalafil and two alpha blockers: doxazosin and tamsulosin. *Int J Imp Res 15* (Suppl 6): 560 (Abst)
14 Kloner RA (2004) Cardiovascular Safety of Phosphodiesterase-5 Inhibitors. In: RA Kloner (ed): *Heart Disease and Erectile Dysfunction*. Humana Press, New Jersey, 139–161
15 Jackson G (2002) Phosphodiesterase type 5 inhibition in cardiovascular disease: experimental models and potential clinical applications. *Eur Heart J* 4, Supplements (Suppl H): H19–H23
16 Langtry HD, Markham A (1999) Sildenafil: A review of its use in erectile dysfunction. *Drugs* 57: 967–989
17 Jackson G (2003) PDE5 inhibitors: looking beyond erectile dysfunction. *Int J Clin Pract* 57: 159–160

Cardiovascular safety of sildenafil in the treatment of erectile dysfunction

Stephan Rosenkranz and Erland Erdmann

Klinik III für Innere Medizin, Universität zu Köln, Joseph-Stelzmann-Str. 9, 50924 Köln (Lindenthal), Germany

Introduction

After sildenafil was launched in 1998, there were numerous reports about deaths that were associated with its use particularly in men with cardiovascular diseases. These reports have raised serious concerns about the cardiovascular safety of PDE5 inhibitors, and several investigations have addressed this issue in the past few years. These studies have concordantly shown that the use of sildenafil is safe in patients with stable cardiac diseases. However, when cardiovascular safety of sildenafil use in clinical practice is considered, several issues have to be taken into account. These include (i) the co-morbidity of erectile dysfunction (ED) and cardiovascular diseases, (ii) the risk of sexual intercourse itself in patients with heart disease, (iii) the efficacy and possible adverse effects of ED treatment with PDE5 inhibitors in patients with specific cardiovascular diseases, (iv) interactions between PDE5 inhibitors and cardiovascular drugs, and (v) the influence of PDE5 inhibitor use on cardiovascular mortality.

Co-morbidity of ED and cardiovascular diseases

The prevalence of both ED and cardiovascular diseases increases with aging. According to the Massachusetts Male Aging Study (MMAS), the prevalence of ED in men between 40 and 70 years of age is 52% [1]. Recent data on sexual function in men >50 years obtained from the Health Professionals Follow-up Study indicated that the lifestyle factors most strongly associated with ED are low physical activity and obesity [2]. This and other studies also revealed that ED and atherosclerotic vascular disease share the same risk factors including age, smoking, hypertension, diabetes, hypercholesterolemia, obesity, and low physical activity. Since these conditions represent important contributors to the pathogenic events involved in both atherosclerosis and ED, it is not surprising that a high degree of co-morbidity of ED and atherosclerotic vascular disease has been reported. Endothelial dysfunction, which represents an early phase of vascular damage, is considered the common denominator [3].

In the MMAS, the prevalence of complete ED in the whole study population was 9.6%, whereas in men with heart disease it was 39% [1]. This study also revealed that in men with cardiac disease, smoking is strongly correlated with ED. In men with chronic stable coronary artery disease, ED is extremely common, affecting approximately 75% of patients [4]. Furthermore, an angiographical study revealed that the severity of ED even correlates with the number of atherosclerotic coronary vessels [5]. In addition to ischemic heart disease, ED is also a frequent problem in patients with congestive heart failure, as 60–70% of heart failure clinic outpatients have been shown to be affected [6]. Since men with ED are likely to have significant cardiac disease and may therefore be at increased risk during physical activity, patients with newly diagnosed ED should always be evaluated for possible coexisting cardiovascular diseases prior to initiation of medical treatment for ED.

Cardiovascular risk associated with sexual activity in patients with heart disease

Medical treatment of ED should only be initiated in men in whom physical activity and emotional stress during sexual intercourse is not associated with an increased risk of cardiovascular events such as myocardial infarction, stroke, or sudden death. Therefore, it is important to consider the cardiovascular risk associated with sexual activity in patients with specified cardiac diseases.

Coronary heart disease

The risk of experiencing a myocardial infarction (MI) during exertion is 10 times greater than the risk at other times, with the highest risk among patients classified as habitually very low active or low active compared to usual physical activity [7]. Hence, the number of MIs expected for men with ED engaged in sexual activity with habitually low activity levels is likely to be greater than those for men in the general population. The exercise workload during sexual activity (3 to 4 METs) is similar to usual daily activities (stair climbing, office work) or mild to moderate intensity exercise [8, 9]. Nevertheless, relevant ischemias may occur during intercourse and have been observed in approximately 50% of non-revascularized post-myocardial infarction patients [10]. Patients at increased risk for myocardial ischemia during sexual activity can reliably be identified by exercise treadmill testing (ETT) [11]. The risk of myocardial infarction during intercourse is slightly elevated. A study on 1.774 patients in 45 US hospitals revealed that sexual activity was a probable trigger of MI in 0.9% of cases [12]. Mortality associated with sexual intercourse is very low, as coital death occurs extremely seldom and accounts for only 0.2–0.6% of sudden death cases [13, 14].

Heart failure

In patients with congestive heart failure, physical exertion may cause cardiac decompensation in a subset of patients. According to the NYHA classification, Class I patients are able to perform daily activities without the occurrence of symptoms despite a diminished left ventricular function. In patients graded NYHA II, particularly in those who are well adjusted by medical treatment, the risk of cardiac decompensation is low. However, in patients at NYHA stages III and IV, who are symptomatic during mild-or-moderate physical exertion or even at rest, sexual activity is likely to ameliorate myocardial function, and the risk of cardiac decompensation is high. Therefore, these patients should first be stabilized before sexual intercourse or ED treatments are engaged [15–17]. In a patient with heart failure NYHA III and secondary pulmonary hypertension who was equipped with a "chronicle device" (which measures hemodynamics in the pulmonary circulation), sexual intercourse was associated with a significant increase of the pulmonary arterial pressure and intermittent atrial fibrillation [18].

Arrhythmias

In addition to myocardial ischemia and cardiac decompensation, complex ventricular arrhythmia may be evoked by physical or emotional exertion. Rhythm disturbances on sexual activity were recorded by 24-h echocardiogram (ECG) monitoring and compared with their occurrence during daily activities in 88 patients with coronary artery disease. In this prospective study, arrhythmias were found in 56% of patients during sexual activity [19]. However, exacerbation of ectopic activity was found in only 11% of patients during intercourse, and complex ventricular arrhythmia was detected in 12.5%. If ventricular ectopic activity occurred on intercourse, it was most often simple and essentially similar to disturbances in daily activity. Hence, there is no significant risk of complex arrhythmias during intercourse in patients with stable coronary artery disease.

Valvular diseases

Patients with severe valvular diseases are considered at increased cardiac risk during sexual intercourse [15–17]. Therefore, patients with a murmur of unknown cause should first undergo specialized evaluation (e.g., echocardiography) before sexual activity or medical ED treatment can be considered safe.

Mild stenosis or regurgitation of the aortic or mitral valve (stage I–II) is not associated with an increased risk of cardiovascular events during mild-to-moderate exertion such as sexual activity. In all other patients sexual intercourse should be deferred until specialist treatment (e.g., valve replacement) has been completed.

The assessment of individual cardiovascular risk is indispensable to optimizing patient safety when contemplating the risk of sexual activity and medical treatment for ED. Since risk stratification based on objective criteria and in particular cases is essential for patient counseling regarding their individual risk, American College of Cardiology (ACC)/American Heart Association (AHA) consensus documents and the Princeton Consensus Panel estimated the risk of sexual activity and ED treatment in patients with specified cardiovascular diseases [15–17], and yielded recommendations for the management of sexual dysfunction in patients with heart disease (Tab. 1).

Table 1. Recommendations for patient counceling regarding sexual activity and ED treatment in patients with cardiovascular disease

Cardiovascular status upon presentation	Recommendations
I. LOW RISK • Minimal/mild stable angina • Asymptomatic, and <3 risk factors for CAD (excluding age and gender) • Controlled hypertension • Mild valvular disease • Post-successful revascularization • Congestive heart failure NYHA I	• Manage within the primary care setting • Sexual activity considered safe • Review treatment options for ED with patient and partner (where necessary)
II. INTERMEDIATE RISK • Moderate, stable angina • Asymptomatic, but >3 risk factors for CAD (excluding age and gender) • Recent MI or CVA (within 2–6 wks.) • Congestive heart failure NYHA II • Murmur of unknown cause • Heart transplant • Recurrent transischemic attacks	• Specialized evaluation recommended (e.g. ETT for angina, echo for murmur) • Patient to be placed into high or low risk category depening on outcome of testing
III. HIGH RISK • Severe or unstable or refractory angina • Uncontrolled hypertension (SBP > 180 mmHg) • Congestive heart failure NYHA III-IV • Recent MI or CVA (<2 weeks) • Complex arrhythmias • Hypertrophic cardiomyopathy • Moderate/severe valvular disease	• Refer for specialized cardiac evaluation and management • Defer sexual activity/ED treatment until cardiac condition established and/or specialist evaluation completed

Treatment of ED with PDE5 inhibitors in patients with cardiovascular diseases

In addition to the cardiac risk associated with physical and emotional exertion during sexual activity, direct effects of PDE5 inhibitors on the cardiovascular system (i.e., coronary hemodynamics, myocardial function, sympathetic activity, arrhythmias) and interactions between sildenafil and cardiovascular drugs may affect the efficacy and safety of medical ED treatment in patients with cardiovascular diseases.

Coronary heart disease

Adverse cardiovascular events associated with sildenafil may be due to myocardial ischemia during sexual activity, and aggravation of ischemia by a vasodilator effect. When the effects of oral sildenafil (100 mg) on coronary hemodynamics were assessed in patients with severe coronary heart disease, there were no significant changes in average peak flow velocity, coronary-artery diameter, volumetric coronary blood flow, coronary vascular resistance, or coronary flow reserve in both diseased and non-diseased coronary arteries [20]. Consistent with these data, sildenafil had no effect on symptoms, exercise duration, or presence or extent of exercise-induced ischemia, as assessed by exercise echocardiography, in men with stable coronary artery disease [21]. Furthermore, a recent study revealed that sildenafil does not reduce exercise tolerance in men with erectile dysfunction and chronic stable angina [22]. These studies collectively indicate that the use of PDE5 inhibitors does not affect coronary blood flow and is safe in patients with stable coronary artery disease.

Heart failure

In the heart, PDE5 is highly expressed in vascular smooth muscle cells of pericardial arteries but not in cardiac myocytes [23]. It is therefore not surprising that sildenafil has no influence on cardiac contractility [24]. In patients with heart failure, sildenafil was shown to increase endothelium-dependent, flow-mediated vasodilation [25], and this effect correlated with improved exercise capacity and a reduction of heart rate during exercise [26], which may even decrease myocardial oxygen consumption during sexual activity. These studies have demonstrated that sildenafil is effective for ED treatment and well tolerated in patients with heart failure.

Arrhythmias

In healthy male volunteers, sildenafil significantly increased sympathetic activity and plasma noradrenalin levels [27]. Furthermore, it reduced vagal modulation and increased sympathetic modulation in patients with heart failure [28]. These autonomic system changes could alter QT dynamics and favour the onset of ventricular arrhythmias. However, at relevant concentrations sildenafil did not affect cardiac repolarization as it did not prolong QT intervals or increase QT dispersion [28, 29]. These data are consistent with a recent animal study demonstrating that sildenafil did not prolong cardiac repolarization at clinically relevant doses, but accelerated cardiac repolarization at supratherapeutic concentrations [30]. In accordance with clinical observations, there is currently no evidence that PDE5 inhibitors predispose patients to malignant ventricular arrhythmias.

Interactions between PDE5 inhibitors and cardiovascular drugs

The target enzyme of sildenafil, phosphodiesterase type 5, is not only present in the corpora cavernosae but also in vascular smooth muscle. Therefore, PDE5 inhibitors such as sildenafil exert vasodilative responses affecting both the arterious and venous system, which are responsible for the typical adverse effects such as headache, flushing, and rhinitis. Studies on human hemodynamics have shown that oral sildenafil leads to a moderate decrease of both systolic (8–10 mmHg) and diastolic (3–6 mmHg) blood pressure, which by itself is not clinically relevant [31, 32]. However, the combination of PDE5 inhibitors and other blood-pressure lowering compounds could potentially cause significant hypotension.

Organic nitrates/NO-donators

The most important contraindication for the treatment of ED with sildenafil is the concomitant use of organic nitrates or other NO donators. Like nitrates, sildenafil causes an NO-mediated accumulation of cyclic guanosin-monophosphate (cGMP) in vascular smooth muscle cells, thereby mediating vasodilatation. However, these effects are not mediated by the same mechanism and may therefore potentiate each other. In healthy male subjects who had taken sildenafil, the intravenous or sublingual administration of glycerol trinitrate led to dramatic decreases of systolic blood pressure which were four times greater than placebo (>25 mmHg) and caused symptomatic hypotension [31]. Hence, sildenafil potentiates the blood pressure-lowering properties of organic nitrates and the combined intake may result in uncontrolled, life-threatening hypotension. Therefore, the coadministration of PDE5 inhibitors and nitrates is strictly contraindicated, and a wash-out period of at least 24 hours is rec-

Table 2. Contraindications for the use of PDE V inhibitors in patients with cardiovascular disease

Absolute contraindications
• Concomitant use of organic nitrates or NO-donators • Recent myocardial infarction (<2 weeks) • Congestive heart failure NYHA III-IV • Unstable angina • Hypertrophic cardiomyopathy • Severe valvular disease • Complex ventricular arrhythmias
Relative contraindications
• Stable angina • Uncontrolled hypertension (SBP > 180 mmHg) • Congestive heart failure NYHA II • Recurrent transischemic attacks • Concomitant use of HIV protease inhibitors

ommended for separation of these drugs [15] (Tab. 2). In ED patients developing angina during sexual activity after sildenafil intake, nitrates must also be avoided, and alternative treatment options include sedation, calcium channel blockers, β-blockers and i.v. fluids [33].

Antihypertensive medication

In contrast to organic nitrates and NO donators, antihypertensive drugs of the various classes (β-blockers, β1-blockers, ACE inhibitors, diuretics, calcium channel blockers) act independently from NO/cGMP. Therefore, the concomitant use of PDE5 inhibitors leads to moderate additive decreases of systolic and diastolic blood pressure, but the effects do not potentiate each other and hence do not result in uncontrolled hypotension [32]. The efficacy of oral sildenafil for ED treatment is comparable in patients taking and those not taking antihypertensive medication [34]. When sildenafil was given to hypertensive patients treated with the calcium channel blocker amlodipine, it caused a mean additive reduction of systolic and diastolic blood pressure of 8 and 7 mmHg, respectively, when compared to placebo (Webb et al., 1999). A *post hoc* analysis of five randomized, placebo-controlled studies involving 1685 men revealed that the effects of oral sildenafil on blood pressure and heart rate in men with ED were small and not clinically relevant in those taking concomitant antihypertensive medication [35]. Furthermore, the incidence of sildenafil-related total adverse events as well as adverse events potentially related to blood pressure decreases (e.g., hypotension, dizziness, syncope) was not different in patients taking or not taking antihypertensive medications [34]. In this study, the number of antihypertensive medications taken from among the five classes (β-blockers, β1-blockers, ACE inhibitors, diuretics, calcium channel blockers) had no effect on the adverse event profile of sildenafil. Taken together, these studies indicate that sildenafil is an effective and well-tolerated treatment for ED in patients taking concomitant antihypertensive medication, including those on multidrug regimens.

The cytochrome P450 system

Sildenafil is metabolized via CYP3A4 and CYP2C9. Therefore, the combined intake of sildenafil and other drugs which are metabolized via CYP3A4 and CYP2C9 – including cardiovascular medications – may result in relevant pharmacokinetic interactions. However, sildenafil did not affect the plasma levels of any other drug studied. Coadministered drugs that inhibit CYP3A4 raise the plasma levels of sildenafil by up to 2–3-fold, mainly by inhibiting first pass metabolism. Since sildenafil is administered prn and has a wide safety margin, these interactions are not considered to be relevant and do not represent a major safety issue. Nevertheless, the coadministration of the HIV protease inhibitor ritonavir with PDE5 inhibitors is not recommended (see Tab. 2).

PDE5 inhibitors and cardiovascular mortality

During the initial four-month period of postmarketing surveillance (March–June 1998), 69 deaths that were associated with the use of sildenafil were reported to the FDA [36]. The majority of these events occurred within four to five hours after taking sildenafil, and death occurred during sexual intercourse (61%) or shortly thereafter. 46 deaths were associated with probable cardiovascular events, and 12 deaths were attributable to possible interactions between sildenafil and organic nitrates [36]. Although the causal relationship between sildenafil intake and death remained unclear in all cases, these reports have raised serious concerns about the cardiovascular safety of PDE5 inhibitors. Consequently, several investigations have addressed this issue in the past few years. These studies have concordantly indicated no evidence for a higher incidence of fatal MI, cerebrovascular events or ischemic heart disease among men taking sildenafil. A prescription event monitoring study performed in 5600 english men revealed that the mortality from ischemic heart disease in sildenafil users was actually 30.1% lower than the expected death rate in men, after adjustment for confounding effects of age [37]. Similarly, no increase in deaths due to myocardial infarction above expected numbers was found in a survey investigating Food & Drug Administration (FDA) reports of deaths in men who had been prescribed sildenafil [38]. Pooled data from 120 clinical trials of sildenafil conducted from 1993–2001 showed that the rates of myocardial infarction and cardiovascular death were low and comparable between men treated with sildenafil or placebo [39]. Preliminary data suggesting a decreased cardiovascular mortality among sildenafil users should thus far be interpreted with caution, as controlled trials investigating these effects for longer observation periods are pending.

In summary, there is no evidence for a higher mortality in sildenafil users compared to the expected death rate in the general population, and – provided that contraindications are considered and the drug is taken under proper conditions (Tabs 1 and 2) – the use of sildenafil for ED can be considered safe in patients with stable cardiac diseases.

Possible cardioprotective effects of PDE5 inhibitors

A number of recent studies indicates that PDE5 inhibitors such as sildenafil, vardenafil, and tadalafil – aside from its vasodilative properties – exert additional effects on the vascular wall and myocardium, which may be cardioprotective. These include favourable effects on endothelial dysfunction, improvement of exercise capacity in patients with heart failure and those with coronary artery disease, beneficial effects in patients with pulmonary hypertension, and a positive influence on peripheral blood flow [40].

Sildenafil significantly improved the impaired endothelial function in patients with Type II diabetes, and this effect persisted for at least 24 h after cessation of chronic therapy, whereas it did not improve NO-mediated endothelium-dependent vascular responses in smokers [41, 42]. In patients with congestive heart failure, improvement of endothelial function by sildenafil correlated with an increase of the exercise capacity and a decrease in heart rate, giving PDE5 inhibition a potential role in the treatment of congestive heart failure [25, 26]. Sildenafil was also shown to dilate epicardial coronary arteries, improve endothelial dysfunction and myocardial ischemia during exercise in patients with coronary artery disease [43]. A relatively large body of evidence indicates that PDE5 inhibitors may be useful in the management of pulmonary hypertension, either alone or as adjunct therapy to inhaled iloprost [44]. The favourable effects on the pulmonary circulation have been shown to correlate with a decrease of a negative prognostic marker in patients with right heart failure, brain natriuretic peptide [45]. Furthermore, recent evidence indicates that PDE5 inhibitors may also improve peripheral blood flow in patients with Raynaud's syndrome [45].

An interesting observation is the fact, that at least some of the effects of PDE5 inhibitors are only seen after chronic treatment and persisted after cessation of therapy. These observations may indicate, that in addition to acute vasodilative effects, chronic treatment with PDE5 inhibitors such as sildenafil may also exert other effects including anti-proliferative, angiogenic, anti-migratory, or anti-fibrotic effects, as already shown in some *in vitro* studies [46, 47]. It will be interesting to learn from future studies, whether these potentially beneficial acute and sustained effects of PDE5 inhibition are clinically relevant and improve the outcome of patients with cardiovascular disease.

References

1 Feldman HA, Goldstein I, Hatzichristou DG, Krane RJ, McKinkey JB (1994) Impotence and its medical and psychological correlates: Results of the Massachusetts Male Aging Study. *J Urol* 151: 54–61
2 Bacon CG, Mittleman MA, Kawachi I, Giovannucci E, Glasser DB, Rimm EB (2003) Sexual function in men older than 50 years of age: Results from the Health Professionals Follow-up Study. *Ann Intern Med* 139: 161–168
3 Solomon H, Man JW, Jackson G (2003) Erectile dysfunction and the cardiovascular patient: endothelial dysfunction is the common denominator. *Heart* 89: 251–253
4 Kloner RA, Mullin SH, Shook T, Matthews R, Mayeda G, Burstein S, Peled H, Poll C, Choudhary R, Rosen R, Padma-Nathan H (2003) Erectile dysfunction in the cardiac patient: How common and should we treat? *J Urol* 170: S46–S50
5 Greenstein A, Chen J, Miller H, Matzkin H, Villa Y, Braf Z (1997) Does severity of ischemic coronary disease correlate with erectile dysfunction? *Int J Impot Res* 9: 123–126
6 Jaarsma T, Dracup K, Walden J, Stevenson LW (1996) Sexual function in patients with advanced heart failure. *Heart Lung* 25: 262–270
7 Giri S, Thompson PD, Kiernan FJ, Clive J, Fram DB, Mitchel JF, Hirst JA, McKay RG, Waters DD (1999) Clinical and angiographic characteristics of exertion-related acute myocardial infarction. *JAMA* 282: 1731–1736
8 Larson JL, McNaughton MW, Kennedy JW, Mansfield LW (1980) Heart rate and blood pressure responses to sexual activity and a stair climbing test. *Heart Lung* 9: 1025–1030
9 Kloner RA, Zusman RM (1999) Cardiovascular effects of sildenafil citrate and recommendations for its use. *Am J Cardiol* 84: 11N–17N
10 Hellerstein HK, Friedman EH (1970) Sexual activity and the postcoronary patient. *Arch Intern Med* 125: 987–999
11 Drory Y, Shapira I, Fisman EZ, Pines A (1995) Myocardial ischemia during sexual activity in patients with coronary artery disease. *Am J Cardiol* 75: 835–837
12 Muller JE, Mittleman MA, MacIure M, Sherwood JB, Tofler GH (1996) Triggering myocardial infarction by sexual activity. *JAMA* 275: 1405–1409
13 Ueno M (1963) The so-called coition death. *Jpn J Leg Med* 127: 333–340
14 Parzeller M, Raschia C, Bratzke H (1999) Der plötzliche kardiovaskuläre Tod bei der sexuellen Betätigung – Ergebnisse einer rechtsmedizinischen Obduktionsstudie. *Z Kardiol* 88: 44–48
15 Cheitlin MD, Hutter AM, Brindis RG, Ganz P, Kaul S, Russell RO, Jr., Zusman RM. ACC/AHA Expert Consensus Document (1999) Use of Sildenafil (Viagra) in patients with cardiovascular disease. *Circulation* 99: 168–177
16 Jackson G, Benjamin N, Jackson N, Allen MJ (1999) Effects of sildenafil citrate on human hemodynamics. *Am J Cardiol* 83 (Suppl. 5A): 13C–20C
17 DeBusk R, Drory Y, Goldstein I, Jackson G, Kaul S, Kimmel SE, Kostis JB, Kloner RA, Lakin M, Meston CM et al. (2000) Management of sexual dysfunction in patients with cardiovascular disease: Recommendations of the Princeton Consensus Panel. *Am J Cardiol* 86: 175–181
18 Cremers B, Kjellström B, Südkamp M, Böhm M (2002) Hemodynamic monitoring during sexual intercourse and physical exercise in a patient with chronic heart failure and pulmonary hypertension. *Am J Med* 112: 428–430
19 Drory Y, Fisman EZ, Shapira I, Pines A (1996) Ventricular arrhythmias during sexual activity in patients with coronary artery disease. *Chest* 109: 922–924
20 Herrmann HC, Chang G, Klugherz BD, Mahoney PD (2000) Hemodynamic effects of sildenafil in men with severe coronary artery disease. *N Engl J Med* 342: 1622–1626
21 Arruda-Olson AM, Mahoney DW, Nehra A, Leckel M, Pellikka PA (2002) Cardiovascular effects of sildenafil during exercise in men with known or probable coronary artery disease. *JAMA* 287: 719–725
22 Fox KM, Thadani U, Ma PTS, Nash SD, Keating Z, Czorniak MA, Gillies H, Keltai M, on Behalf

of the CAESAR I (Clinical American and European Studies of Angina and Revascularization) Investigators (2003) Sildenafil citrate does not reduce exercise tolerance in men with erectile dysfunction and chronic stable angina. *Eur Heart J* 24: 2206–2212
23 Parums DV, Charleton R, Johnson N, Cindrova T, Skepper J, Phillips S, Ridden J, Burslem F (2000) Immunohistochemical (ICH), *in situ* hybridisation (ISH) and biochemical characterisation of phosphodiesterase type 5 (PDE5) in normal and ischeamic human cardiac tissue. *Eur Heart J* 21 (Suppl.): 616 (Abstract)
24 Cremers B, Scheler M, Maack C, Gröschel A, Schäfers HJ, Böhm M (2001) Einfluß von Sildenafil (Viagra) auf die kardiale Kontraktilität und den Gefäßtonus menschlicher Arteriae mammariae und Venae saphenae magnae. *Z Kardiol* 90 (Suppl. 2): 208 (Abstract)
25 Katz SD, Balidemaj K, Homma S, Wu H, Wang J, Maybaum S (2000) Acute type 5 phosphodiesterase inhibition with sildenafil enhances flow-mediated vasodilation in patients with chronic heart failure. *J Am Coll Cardiol* 36: 845–851
26 Bocchi EA, Guimaraes G, Mocelin A, Bacal F, Bellotti G, Ramires JF (2002) Sildenafil effects on exercise, neurohormonal activation, and erectile dysfunction in congestive heart failure. *Circulation* 106: 1097–1103
27 Phillips BG, Kato M, Pesek CA, Winnicki M, Narkiewicz K, Davison D, Somers VK (2000) Sympathetic activation by sildenafil. *Circulation* 102: 3068–3073
28 Piccirillo G, Nocco M, Lionetti M, Moise A, Naso C, Marigliano V, Cacciafesta M (2002) Effects of sildenafil citrate (Viagra) on cardiac repolarization and on autonomic control in subjects with chronic heart failure. *Am Heart J* 143: 703–710
29 Alpaslan M, Onrat E, Samli M, Dincel C (2003) Sildenafil citrate does not affect QT intervals or QT dispersion: An important observation for drug safety. *Ann Noninvasive Electrocardiol* 8: 14–17
30 Chiang C-E, Luk H-N, Wang T-M, Ding PYA (2002) Effects of sildenafil on cardiac repolarization. *Cardiovasc Res* 55: 290–299
31 Webb DJ, Freestone S, Allen MJ, Muirhead GJ (1999) Sildenafil citrate and blood-pressure lowering drugs: Results of drug interaction studies with an organic nitrate and a calcium antagonist. *Am J Cardiol* 83 (Suppl. 5A): 21C–28C
32 Rosenkranz S, Erdmann E (2001) Wechselwirkungen zwischen Sildenafil und Antihypertensiva – was ist gesichert? *Dtsch Med Wschr* 126: 1144–1149
33 Dorsch A, Erdmann E, Hempelmann G, Kupper W, Maisch B, Maurer G (1998) Therapieempfehlung "Viagra und NO-DonatoreN". *Herz/Kreisl* 30: XI
34 Kloner RA, Brown M, Prisant LM, Collins M (2001) Effect of sildenafil in patients with erectile dysfunction taking antihypertensive therapy. Sildenafil Study Group. *Am J Hypertens* 14: 70–73
35 Zusman RM, Prisant LM, Brown MJ (2000) Effect of sildenafil citrate on blood pressure and heart rate in men with erectile dysfunction taking concomitant antihypertensive medication. Sildenafil Study Group. *J Hypertens* 18: 1865–1869
36 Centre for Drug Evaluation and Research (1998) Food and Drug Administration. Post marketing safety of sildenafil citrate (Viagra): Summary of reports of death in Viagra users received from marketing (late March to June 1998)
37 Shakir SAW, Wilton LV, Boshier A, Layton D, Heeley E (2001) Cardiovascular events in users of sildenafil: results from first phase of prescription event monitoring in England. *BMJ* 322: 651–652
38 Wysowsky DK, Farinas E, Swartz L (2002) Comparison of reported deaths and expected deaths in sildenafil (Viagra) users. *Am J Cardiol* 89: 1331–1334
39 Mittleman MA, Glasser DB, Orazem J (2003) Clinical trials of sildenafil citrate (Viagra) demonstrate no increase in the risk of myocardial infarction and cardiovascular death compared with placebo. *Int J Clin Pract* 57: 597–600
40 Cremers B, Böhm M (2003) Non erectile dysfunction application of sildenafil. *Herz* 28: 325–333
41 Desouza C, Parulkar A, Lumpkin D, Akers D, Fonseca VA (2002) Acute and prolonged effects of sildenafil on brachial artery flow-mediated dilatation in type 2 diabetes. *Diabetes Care* 25: 1336–1339
42 Dishy V, Harris PA, Pierce R, Prasad HC, Sofowora G, Bonar HL, Wood AJJ, Stein CM (2003) Sildenafil does not improve nitric oxide-mediated endothelium-dependent vascular responses in smokers. *Br J Clin Pharmacol* 57: 209–212
43 Halcox JP, Nour KR, Zalos G, Mincemoyer RA, Waclawiw M, Rivera CE, Willie G, Ellaham S, Quyyumi AA (2002) The effect of sildenafil on human vascular function, platelet activation, and myocardial ischemia. *J Am Coll Cardiol* 40: 1232–1240

44 Ghofrani HA, Olschewski H, Seeger W, Grimminger F (2002) Sildenafil for treatment of severe pulmonary hypertension and commencing right heart failure. *Pneumol* 56: 665–672
45 Rosenkranz S, Diet F, Karasch T, Weihrauch J, Wassermann K, Erdmann E (2003) Sildenafil improved pulmonary hypertension and peripheral blood flow in a patient with scleroderma-associated lung fibrosis and the Raynaud phenomenon. *Ann Intern Med* 139: 9–10
46 Osinski MT, Rauch BH, Schrör K (2001) Antimitogenic actions of organic nitrates are potentiated by sildenafil and mediated via activation of protein kinase A. *Mol Pharmacol* 59: 1044–1050
47 Zhang R, Wang L, Zhang L, Chen J, Zhu Z, Zhang Z, Chopp M (2003) Nitric oxide enhances angiogenesis via the synthesis of vascular endothelial growth factor and cGMP after stroke in the rat. *Circ Res* 92: 308–313

NO pathway and phosphodiesterase inhibitors in pulmonary arterial hypertension

Hossein Ardeschir Ghofrani, Werner Seeger and Friedrich Grimminger

Department of Internal Medicine, Pulmonary Hypertension Center, University Hospital, Klinikstrasse 36, 35392 Giessen, Germany

Signaling pathway of nitric oxide and phosphodiesterase in pulmonary hypertension

Nitric oxide (NO) is constitutively produced in the lung by NO-synthases. The main cellular sources of lung NO production are the vascular endothelium and the airway epithelia [1, 2]. Adaptation of the perfusion distribution to well ventilated areas of the lung (ventilation/perfusion (V/Q) matching) is mainly regulated by local NO production [3, 4]. NO- synthase activity is regulated on transcriptional and post-translational redox-based modulation level [5]. The common signaling pathway of endogenous vasodilators, such as nitric oxide, prostaglandins, and natriuretic peptides, engage cyclic nucleotides (cAMP and cGMP). The enzymatic source of these second messengers are mainly adenylate- and guanylate-cyclases [6]. PDEs represent a superfamily of enzymes, with PDE1 through PDE11 being currently known, that inactivate cyclic AMP and cyclic GMP, with different tissue distribution and substrate specificities [6, 7]. Depending on their selective profile, PDE inhibitors differentially regulate the activity of cAMP and/or cGMP. Thus, they might be offered as therapeutic tools to augment and prolong prostanoid- and NO-related vascular effects. The efficacy of this approach has been proven in various experimental studies [8, 9]. The most important cyclic GMP degrading phosphodiesterase – PDE5 – is abundantly expressed in lung tissue [7]. The selective PDE5 inhibitor sildenafil has been approved for the treatment of erectile dysfunction [10]. Documented use in numerous otherwise healthy individuals and patients with a variety of underlying diseases sildenafil displayed an excellent safety profile [11].

Established treatments for pulmonary arterial hypertension

Currently, continuous infusion of prostacyclin is the most established treatment of severe pulmonary arterial hypertension, since this approach has been shown to be life-saving [12] and to improve exercise capacity [13] in controlled trials.

However, there are also drawbacks of this therapy immanent to the drug and to the necessity of implanting a continuous intravenous line. These undesired effects include substantial systemic side effects due to lack of pulmonary selectivity, the need of progressive dosage increase, and septic complications of the intravenous line. In order to preserve the advantageous effects of prostacyclin, and to avoid several of the previously mentioned side effects, the concept of aerosolized iloprost (long acting prostacyclin analogue) for treatment of pulmonary arterial hypertension (PAH) was developed [14, 15]. The results of a double-blind placebo-controlled multicenter study just recently demonstrated that daily inhaled iloprost significantly improved exercise capacity, NYHA classification, dyspnoea scoring and event-free survival over a three month observation period in patients with selected forms of PAH and chronic thromboembolic pulmonary hypertension [16]. A new promising agent for medical treatment of PAH is the nonselective oral endothelin receptor antagonist bosentan. In a controlled Phase III study, bosentan showed beneficial effects on exercise tolerance in patients with primary pulmonary hypertension (PPH) and those with pulmonary hypertension associated with collagen vascular disease [17]. As the most prominent side effect, liver toxicity was documented in a considerable percentage of patients. Long-term experience in chronically treated patients will give insight into to occurrence this complication in clinical use.

The search goes on for an "ideal" pulmonary vasodilator that combines pulmonary selectivity with simplicity of administration and reduced side effects. Recently, the phosphodiesterase 5 inhibitor sildenafil has come into the focus of investigation.

Sildenafil in experimental pulmonary hypertension

In animal experiments, several PDE inhibitors displayed favorable pulmonary vasodilatory potential [8, 18, 19]. Sildenafil in such setting proved to be a potent and pulmonary selective vasodilator [20]. Most interestingly, this agent was also able to reduce hypoxia induced pulmonary hypertension in man and in an experimental animal model [21]. The effects of sildenafil on chronic remodeling processes in the pulmonary vasculature are not yet well known. Future results derived from long term experimental models of pulmonary hypertension will hopefully give insight into these mechanisms.

Clinical experience with sildenafil for the treatment of chronic pulmonary hypertension

The vasodilatory effects of nitric oxide administered by inhalation are restricted to the pulmonary vasculature. Nitric oxide has a very short half-life, is used as a screening agent for lung vasoreactivity [22], and is effective for improving gas exchange in selected patients with the adult respiratory distress syndrome (ARDS) [23]. Weaning from chronic nitric oxide treatment in patients with the adult respiratory distress syndrome was found to be facilitated by oral sildenafil [24]. In patients with pulmonary arterial hypertension, short-term application of sildenafil during right heart catheterization effectively reduced pulmonary vascular resistance in a dose dependent manner. Notably, the vasodilatory effects were mainly restricted to the pulmonary circulation and were significantly stronger than the effects seen with inhaled nitric oxide [25]. In combination with inhaled iloprost, augmentation of the pulmonary vasodilatory effect of each single agent was noted [25, 26]. In patients with deteriorating severe PAH despite ongoing prostanoid treatment, additional long-term administration of oral sildenafil improved exercise capacity and pulmonary hemodynamics [27]. The combination of prostanoids and sildenafil has potential as possible future treatment for pulmonary hypertension. Numerous reports about the clinical use of sildenafil in pulmonary arterial hypertension in uncontrolled trials have been published so far [25, 28–35].

Interestingly, sildenafil appears to be effective for treating patients with pulmonary hypertension of other origin than primary pulmonary hypertension. In patients suffering from HIV-related pulmonary hypertension, sildenafil was similarly effective in reducing pulmonary vascular resistance as in PPH [36]. Moreover, this therapeutic approach has been reported to be effective in pediatric patients [37]. When pulmonary hypertension is associated with interstitial lung disease, systemic administration of vasodilators redundantly increases the blood flow to low or non-ventilated lung areas by interfering with the physiological hypoxic vasoconstrictor mechanism, thereby worsening preexistent ventilation(V)/perfusion(Q) mismatch and shunt flow. The decrease in arterial oxygenation and wasting of the small ventilatory reserve of these patients are the negative consequences of this effect. Most interestingly, oral sildenafil was found to cause pulmonary vasodilatation in patients with lung fibrosis and pulmonary hypertension, with the overall vasodilatory potency corresponding to that of intravenous prostacyclin. In contrast to the infused prostanoid, selectivity for well-ventilated lung areas was demonstrated for sildenafil, resulting in an improvement rather than deterioration of gas exchange [38]. In addition, recent data suggest beneficial long term effectiveness in patients with non-operable chronic thromboembolic pulmonary hypertension [39]. The importance of this finding derives from the fact that there is little to offer as therapeutic option for these patients except lung transplantation.

Limitation of exercise capacity due to hypoxic pulmonary hypertension: impact of sildenafil

Pulmonary hypertension impairs right ventricular performance due to increased right heart afterload. However, it is still unclear to which extent exercise tolerance is limited by this mechanism. In a recent investigation we addressed this issue under conditions of acute hypoxia at sea level and prolonged hypoxia at the altitude of Mount Everest Base Camp [40]. The investigations were performed in healthy volunteers to exclude other confining factors which may add to the limitation of exercise tolerance in patients suffering from chronic hypoxia (e.g., muscle wasting, chronic immobilization, etc.). In essence, acute and prolonged hypoxia induced significant pulmonary hypertension in the study subjects. As expected, exercise tolerance was dramatically reduced as a result of the combination of severe hypoxemia and significant pulmonary hypertension. Sildenafil in this setting significantly reduced pulmonary hypertension under resting conditions as well as during exercise. The most interesting finding of this study was that the reversal of pulmonary hypertension resulted in an immediate improvement of exercise tolerance, irrespective of improvements in oxygenation. Therefore, the results of this study might be stimulatory for investigations addressing the therapeutic potential of this drug in patients suffering from chronic hypoxic pulmonary hypertension as it occurs in various chronic diseases (e.g., chronic obstructive lung disease, interstitial lung disease, obstructive sleep apnea, etc.).

Conclusion

The NO/cGMP axis represents a pivotal signaling pathway for the lung circulation. Phosphodiesterases, as regulators of the second messenger response to endogenous NO, are thus of great therapeutic potential for the treatment of lung circulatory disorders. Among the clinically available PDE-inhibitors, sildenafil is a most promising agent for pulmonary vasodilatation and long-term anti-remodeling in the lung vasculature of PAH patients. Although orally administered, sildenafil does possess features of pulmonary selectivity. It may be favorably combined with other vasodilative and anti-proliferative agents. Controlled clinical trials in patients with pulmonary hypertension of various origin are currently underway and recommendations about the therapeutic use of sildenafil in this area should depend on the results of these studies.

References

1 Bohle RM, Hartmann E, Kinfe T, Ermert L, Seeger W, Fink L (2000) Cell type-specific mRNA quantitation in non-neoplastic tissues after laser-assisted cell picking. *Pathobiology* 68: 191–195
2 German Z, Chambliss KL, Pace MC, Arnet UA, Lowenstein CJ, Shaul PW (2000) Molecular basis of cell-specific endothelial nitric-oxide synthase expression in airway epithelium. *J Biol Chem* 275: 8183–8189
3 Ide H, Nakano H, Ogasa T, Osanai S, Kikuchi K, Iwamoto J (1999) Regulation of pulmonary circulation by alveolar oxygen tension via airway nitric oxide. *J Appl Physiol* 87: 1629–1636
4 Grimminger F, Spriestersbach R, Weissmann N, Walmrath D, Seeger W (1995) Nitric oxide generation and hypoxic vasoconstriction in buffer-perfused rabbit lungs. *J Appl Physiol* 78: 1509–1515
5 Michelakis ED (2003) The role of the NO axis and its therapeutic implications in pulmonary arterial hypertension. *Heart Fail Rev* 8: 5–21

6 Beavo JA (1995) Cyclic nucleotide phosphodiesterases: functional implications of multiple isoforms. *Physiol Rev* 75: 725–748
7 Ahn HS, Foster M, Cable M, Pitts BJ, Sybertz EJ (1991) Ca/CaM-stimulated and cGMP-specific phosphodiesterases in vascular and non-vascular tissues. *Adv Exp Med Biol* 308: 191–197
8 Schermuly RT, Krupnik E, Tenor H, Schudt C, Weissmann N, Rose F, Grimminger F, Seeger W, Walmrath D, Ghofrani HA (2001) Coaerosolization of Phosphodiesterase Inhibitors Markedly Enhances the Pulmonary Vasodilatory Response to Inhaled Iloprost in Experimental Pulmonary Hypertension. Maintenance of lung selectivity. *Am J Respir Crit Care Med* 164: 1694–1700
9 Weimann J, Ullrich R, Hromi J, Fujino Y, Clark MW, Bloch KD, Zapol WM (2000) Sildenafil is a pulmonary vasodilator in awake lambs with acute pulmonary hypertension. *Anesthesiology* 92: 1702–1712
10 Cheitlin MD, Hutter AM, Jr., Brindis RG, Ganz P, Kaul S, Russell RO, Jr., Zusman RM (1999) Use of sildenafil (Viagra) in patients with cardiovascular disease. Technology and Practice Executive Committee. *Circulation* 99: 168–177
11 Padma-Nathan H, Eardley I, Kloner RA, Laties AM, Montorsi F (2002) A 4-year update on the safety of sildenafil citrate (Viagra). *Urology* 60: 67–90
12 Barst RJ, Rubin LJ, Long WA, McGoon MD, Rich S, Badesch DB, Groves BM, Tapson VF, Bourge RC, Brundage BH et al. (1996) A comparison of continuous intravenous epoprostenol (prostacyclin) with conventional therapy for primary pulmonary hypertension. The Primary Pulmonary Hypertension Study Group. *N Engl J Med* 334: 296–302
13 Badesch DB, Tapson VF, McGoon MD, Brundage BH, Rubin LJ, Wigley FM, Rich S, Barst RJ, Barrett PS, Kral KM et al. (2000) Continuous intravenous epoprostenol for pulmonary hypertension due to the scleroderma spectrum of disease. A randomized, controlled trial. *Ann Intern Med* 132: 425–434
14 Olschewski H, Walmrath D, Schermuly R, Ghofrani A, Grimminger F, Seeger W (1996) Aerosolized prostacyclin and iloprost in severe pulmonary hypertension. *Ann Intern Med* 124: 820–824
15 Olschewski H, Ghofrani HA, Schmehl T, Winkler J, Wilkens H, Hoper MM, Behr J, Kleber FX, Seeger W (2000) Inhaled iloprost to treat severe pulmonary hypertension. An uncontrolled trial. German PPH Study Group. *Ann Intern Med* 132: 435–443
16 Olschewski H, G Simonneau, N Galie, T Higenbottam, R Naeije, LJ Rubin, S Nikkho, R Speich, MM Hoeper, J Behr et al. (2002) Inhaled iloprost for severe pulmonary hypertension. *N Engl J Med* 347: 322–329
17 Rubin LJ, DB Badesch, RJ Barst, N Galie, CM Black, A Keogh, T Pulido, A Frost, S Roux, I Leconte et al. (2002) Bosentan therapy for pulmonary arterial hypertension. *N Engl J Med* 346: 896–903
18 Schermuly RT, Roehl A, Weissmann N, Ghofrani HA, Schudt C, Tenor H, Grimminger F, Seeger W, Walmrath D (2000) Subthreshold doses of specific phosphodiesterase type 3 and 4 inhibitors enhance the pulmonary vasodilatory response to nebulized prostacyclin with improvement in gas exchange. *J Pharmacol Exp Ther* 292: 512–520
19 Ichinose F, Adrie C, Hurford WE, Bloch KD, Zapol WM (1998) Selective pulmonary vasodilation induced by aerosolized zaprinast. *Anesthesiology* 88: 410–416
20 Schermuly RT, Kreissleimeier KP, Ghofrani HA, Yilmaz H, Butrous G, Emert L, Emert M, Weissmann N, Rose F, Guenther A et al. (2004) Chronic sildenafil treatment inhibits monocrotaline-induced pulmonary hypertension in rats. *Am J Respir Crit Care Med* 169: 39–45
21 Zhao L, Mason NA, Morrell NW, Kojonazarov B, Sadykov A, Maripov A, Mirrakhimov MM, Aldashev A, Wilkins MR (2001) Sildenafil inhibits hypoxia-induced pulmonary hypertension. *Circulation* 104: 424–428
22 Sitbon O, Humbert M, Jagot JL, Taravella O, Fartoukh M, Parent F, Herve P, Simonneau G (1998) Inhaled nitric oxide as a screening agent for safely identifying responders to oral calcium-channel blockers in primary pulmonary hypertension. *Eur Respir J* 12: 265–270
23 Rossaint R, Falke KJ, Lopez F, Slama K, Pison U, Zapol WM (1993) Inhaled nitric oxide for the adult respiratory distress syndrome. *N Engl J Med* 328: 399–405
24 Atz AM, Wessel DL (1999) Sildenafil ameliorates effects of inhaled nitric oxide withdrawal. *Anesthesiology* 91: 307–310
25 Ghofrani HA, Wiedemann R, Rose F, Olschewski H, Schermuly RT, Weissmann N, Seeger W, Grimminger F (2002) Combination therapy with oral sildenafil and inhaled iloprost for severe pulmonary hypertension. *Ann Intern Med* 136: 515–522

26 Wilkens H, Guth A, Konig J, Forestier N, Cremers B, Hennen B, Bohm M, Sybrecht GW (2001) Effect of inhaled iloprost plus oral sildenafil in patients with primary pulmonary hypertension. *Circulation* 104: 1218–1222

27 Ghofrani HA, Rose F, Schermuly RT, Olschewski H, Wiedemann R, Kreckel A, Weissmann N, Ghofrani S, Enke B, Seeger W, Grimminger F (2003) Oral sildenafil as long-term adjunct therapy to inhaled iloprost in severe pulmonary arterial hypertension. *J Am Coll Cardiol* 42: 158–164

28 Watanabe H, Ohashi K, Takeuchi K, Yamashita K, Yokoyama T, Tran QK, Satoh H, Terada H, Ohashi H, Hayashi H (2002) Sildenafil for primary and secondary pulmonary hypertension. *Clin Pharmacol Ther* 71: 398–402

29 Michelakis E, Tymchak W, Lien D, Webster L, Hashimoto K, Archer S (2002) Oral sildenafil is an effective and specific pulmonary vasodilator in patients with pulmonary arterial hypertension: comparison with inhaled nitric oxide. *Circulation* 105: 2398–2403

30 Zimmermann AT, Calvert AF, Veitch EM (2002) Sildenafil improves right-ventricular parameters and quality of life in primary pulmonary hypertension. *Intern Med J* 32: 424–426

31 Singh B, Gupta R, Punj V, Ghose T, Sapra R, Grover DN, Kaul U (2002) Sildenafil in the management of primary pulmonary hypertension. *Indian Heart J* 54: 297–300

32 Lepore JJ, Maroo A, Pereira NL, Ginns LC, Dec GW, Zapol WM, Bloch KD, Semigran MJ (2002) Effect of sildenafil on the acute pulmonary vasodilator response to inhaled nitric oxide in adults with primary pulmonary hypertension. *Am J Cardiol* 90: 677–680

33 Kothari SS, Duggal B (2002) Chronic oral sildenafil therapy in severe pulmonary artery hypertension. *Indian Heart J* 54: 404–409

34 Sastry BK, Narasimhan C, Reddy NK, Anand B, Prakash GS, Raju PR, Kumar DN (2002) A study of clinical efficacy of sildenafil in patients with primary pulmonary hypertension. *Indian Heart J* 54: 410–414

35 Zhao L, Mason NA, Strange JW, Walker H, Wilkins MR (2003) Beneficial effects of phosphodiesterase 5 inhibition in pulmonary hypertension are influenced by natriuretic Peptide activity. *Circulation* 107: 234–237

36 Carlsen J, Kjeldsen K, Gerstoft J (2002) Sildenafil as a successful treatment of otherwise fatal HIV-related pulmonary hypertension. *AIDS* 16: 1568–1569

37 Abrams D, Schulze-Neick I, Magee AG (2000) Sildenafil as a selective pulmonary vasodilator in childhood primary pulmonary hypertension. *Heart* 84: E4

38 Ghofrani HA, Wiedemann R, Rose F, Schermuly RT, Olschewski H, Weissmann N, Gunther A, Walmrath D, Seeger W, Grimminger F (2002) Sildenafil for treatment of lung fibrosis and pulmonary hypertension: a randomised controlled trial. *Lancet* 360: 895–900

39 Ghofrani HA, Schermuly RT, Rose F, Wiedemann R, Kohstall MG, Kreckel A, Olschewski H, Weissmann N, Enke B, Ghofrani S, Seeger W, Grimminger F (2003) Sildenafil for long-term treatment of nonoperable chronic thromboembolic pulmonary hypertension. *Am J Respir Crit Care Med* 167: 1139–1141

40 Ghofrani HA, Reichenberger F, Kohstall MG, Mrosek E, Seeger T, Seeger W, Grimminger F (2004) Sildenafil increased exercise capacity during hypoxia at low altitudes and at Mount Everest Base Camp: a randomized, double-blind, placebo-controlled crossover trial. *Ann Intern Med* 141: 169–177

The phosphodiesterase V inhibitor low responder study (PILRS) in patients with erectile dysfunction – A rationale for a PDE5 inhibitor combination therapy

Udo Dunzendorfer[1], Arne Behm[2], Eva Dunzendorfer[3] and Annette Dunzendorfer[4]

[1] *J.-W.-G.-University; Zeil 65–69, 60313 Frankfurt/Main, Germany*
[2] *Department of Urology, University of Lübeck Medical School, Ratzeburger Allee 160, 23538 Lübeck, Germany*
[3] *Alois Eckertstr. 28, 60528 Frankfurt/Main, Germany*
[4] *Evangelisches Krankenhaus Ludwigsfelde-Teltow, Akademisches Lehrkrankenhaus der Freien Universität Berlin, Innere Medizin, Albert-Schweitzer-Str. 40–44, 14974 Ludwigsfelde, Germany*

Introduction

Population-based surveys have shwon that erectile dysfunction (ED) is a constant and common condition in aging men [1–3]. The majority experiences mild ED, while moderate or severe ED occurs in up to 26% of the male population [2]. Under pathological situations, such as diabetes, hypertension, cardiovascular diseases, depression, renal and cancer diseases the percentage of occurrence may increase [2, 4–6]. Multiple psychological and organic conditions are involved in the etiology of ED. The current focus on the molecular mechanisms of penile erection is leading to new treatment approaches and analysing the failure of non-responders with respect to drug target mechanisms.

The introduction of a new class of oral agents, the phosphodiesterase 5 (PDE5) inhibitors, such as the protagonist sildenafil, has revolutionised the treatment for ED worldwide, and has also completely changed the prevalence of ED and the desire for improved sexual potency, based upon prescription sales of sildenafil (1998 – 7520×10^3 and 2001 – 14.027×10^3 [7]) *versus* Alprostadil.

The availability of sildenafil led to a substantial increase in men being diagnosed with ED, compared with the rates prior to release of this drug, available data showing 250% increase alone in the US [6].

A second generation of PDE5 agents and other newer drugs will enhance the understanding of the physiologic mechanisms of ED, focused towards the molecular targeting, and will also clarify the drug interaction with target factors influenced by polymorphism [8]. This will help to elucidate why 30–40% of patients are prevalent non-responders representing a major problem in the therapy of ED.

Epidemiology and etiology of non-responders

The exact number of prevalence of patients resistent to therapy with PDE5 inhibitors is difficult to asses. The epidemiology of non-responders is better understood in the light of the epidemiology of untreated ED. The number of non-responders to any single drug treatment targeting penile pharmacology in ED will significantly increase; based on the Massachusetts Male Aging Study, presenting 52% of 1700 men aged 40–70 years surveyed reporting some degree of ED [5]. The estimated incidence of ED almost doubles with each decade of life [3], and in the US the annual incidence of ED for white men between 40 and 70 years is approximately 617,700 [5], 30–40% of them being potential non-responders in case of single drug approach. Similar numbers of non-responders can be estimated when taking into account the annual incidence of ED to 12 per 1000 in the age onto 40–49 years and 46 per 1000 for those aged 60–69 years [2]. The Cologne Male Survey [1] reported a 19.2% prevalence of ED in men aged 30–80 years, with a steep increase in ED associated with age and a further increase of ED in men with specific chronic health problems such as diabetes mellitus, lower urinary tract symptoms, hypertension, and those who had undergone pelvic surgery [1]. The Multi-National Survey of the Aging Male (MSAM-7), a study of 14,000 men aged 50–80 years in seven countries (US, UK, France, Germany, Italy, Spain, and The Netherlands) [9], concluded that high blood pressure is directly associated with the severity of sexual problems. This leads to the assumption that antihypertensive therapy could directly worsen penile influx and efflux, increasing the number of non-responders to patients with PDE5 inhibitors.

Pathophysiology of non-responders

Although there is a small number of reports on patients non-responsive to the treatment of ED, current studies indicate that up to 85% of the causes of ED are of organic nature [10–12]. Vascular diseases, either poor arterial flow (arteriogenic or influx ED) or poor cavernosal trapping of penile blood (veno-occlusive or efflux ED) represent the major cause of organic ED in men aged 40 years and older [5]. The common type of vascular ED occurs as a mixed vascular disease with a combination of veno-occlusive and arterial insufficiency, limiting the relaxation of the corporeal smooth muscle, as well as the dilatation of penile sinusoidal system [5].

Frequently disregarded is the pathophysiology of the *defective regional venous flow* from the corpora cavernosa, influenced by hormonal conditions, such as low testosterone levels that may interfere with androgen effect on the nitric oxide pathway in veins, and neurologic injuries particularly from spinal cord injury or non-nerve-sparing radical prostatectomy [5].

The group of PDE5 inhibitor non-responders may also increase by therapy modalities interfering with cAMP, NO content of the veins and the polymorphism of G Protein β3 subunit [8].

Sexual activity with its multiple regulatory mechanisms can only be artificially separated into various single process units, such as cavernosal, regional, multivarious, and central, interfering with self-supporting and self-enhancing procedures, which collectively contribute to sexual activity. Therefore, it appears biologically contradictory that catecholamines in the central nervous system have stimulatory effects on sexual behaviour; however when injected into the penis they reduce blood influx. On the other hand, it is known that norepinephrine levels increase during sexual activity, enforcing the feeling of orgasm and physiologically preparing the end of the orgasm by reversing blood flow in the corpora cavernosa and thus preventing priapism. The synchronization of the different steps starting with sexual arousal, increase of NO, and ending with relaxation of smooth muscle light chain kinase [7] is highly complicated. Theoretically any of these steps could be affected leading to dysfunction.

Diagnosis of non-responders

Diagnostic features are made of clinical and laboratory assessements. It is not possible to identify non-responders prior to monotherapy with PDE5 inhibitor by a specific test although injection of prostaglandin (10 µg alprostadil intracavernosally) is advocated in the algorithmic diagnostic tool of ED. A negative response to the injection of alprostadil suggests a limited reponse rate to PDE5 inhitors. This test procedure of intracavernosal injection is however not always accepted by the patients starting with an oral drug of their ED. Hormonal analyses do not show significant results in this clinical problem, showing the difficulties of assessment.

In a particular patient, diagnosis of ED may require the analysis of medical risk factors such as diabetes, hypertension, hypercholesterolemia, cardiovascular disease, spinal cord injury, peripheral vascular disease, endocrine abnormalities, pelvic surgery, depression, arthritis, renal failure, trauma, cancer, lifestyle risk factors that include smoking, alcohol, drug use, and obesity, and usage of prescription drugs including beta-blockers, antidepressants, cytotoxic compounds and thiazide diuretics [5]. These medical risk factors however will not precisely predict the failure rate to oral drug therapy.

Multivariate risk factors in elderly men will lead to a higher incidence rate of non-responders. It appears that elderly patients under mixed vascular ED, multiple drug regimens for hypertension, patients after prostatectomy, pelvic surgery, and venous diseases have also a high failure rate for a single-drug approach. Simply a negative response to PDE5 inhibitors following dose titration is also diagnostic for non-responders.

Combinatorial approaches in the treatment of non-responders

To avoid failures in the treatment of ED with single agents especially under multivariate risk conditions in elderly patients this has led to the development of new approaches in treating ED with a combination therapy, where one of the compounds was a PDE5 inhibitor. In this respect, combinations of sildenafil with apomorphine, x-adrenoceptor (x-AR) antagonists, nitrosylated x-AR antagonists have been investigated [9]. Currently new targets such as melanocortin receptors, oxytocin, oxytocin receptors, growth hormone-releasing peptide receptors, 5-hydroxytryptamine receptors [11] as well as guanylyl cyclase activators and RhoA-kinase gene therapy [12] are under investigation to develop effective drugs for the treatment of ED.

The random control of some of these potential drug therapies by the PILRS study showed that most of the drugs mentioned above do not have any synergistic or additional clinical effects in ED-patients, while a few were found to have unexpected synergistic effects, probably by targeting the defective venous system. Only sympathicomimetic compounds were found to improve the PDE5 inhibitor drug effect in ED low- to non-responders [13]. The biology of erection must be understood as a fully integrated procedure based on primary changes in circulation of the corpora cavernosa, as well as on secondary circulatory changes in the urethra, prostate and pelvic region, resulting in the integrated erection status (IES). Therapeutic modulators must be reactive to these primary and secondary targets in ED.

Design of the study

Patients with a constant failure rate to organic ED treatment were identified by a routine clinical examination and admitted to the clinical study. Based on earlier findings they were divided into subgroups A and B starting with different treatment modules such a 50 mg sildenafil or 50 mg sildenafil combined with 5 mg DHE, respectively. In the therapeutic cycle of 28 days the patients then received increased doses of 100 mg sildenafil alone or combined with 5 mg DHE (group B). Following a washout period of 28 days the crossover design was applied, starting with 50 mg sildenafil combined with 5 mg DHE in group A and with 50 mg sildenafil in group B. The co-medication was not changed and the results were categorised according to the IEEF and Wellness Scores. The PILRS investigation was aimed to identify patients sensitive to an additional study impact situation. To keep a routine procedure, all drugs investigated were prescribed and administered within dose range.

The PILRS study is an open non-blinded study derived from a clinical routine standard therapy with drugs prescribed. After the run in period the dose of sildenafil was in the range of 50–100 mg using the therapeutic recommended drug level. The dose of DHE as second drug was given in a 5 mg DHE formulation and in one cycle DHE was also used as drug alone.

Validity analysis of the PILRS investigation

The valid study population was represented by non-responders under therapy with PDE5 inhibitors. The suitability of the group for the clinical study was based on preselected major criteria. Other screening criteria were of interest to patients to learn more about their own sexual reaction, or more precisely to learn more about the unresponsiveness towards PDE5 inhibitors under different conditions, such as additional therapy. The demographic characteristic is consistent with that of an open urological ambulance with patients mostly suffering from benign prostate hyperplasia, all types of diabetes and hypertension. Exclusion criteria included patients treated with nitrate compounds for heart diseases.

The valid treatment for non-responders was the medical guidance during the routine administration by well-known drugs. The drugs orally administered to those patients visiting the office on Monday or Thursday starting with sildenafil 50 mg; after four weeks the dose was increased to 100 mg. Those patients coming in on Tuesday or Friday started with sildenafil 50 mg after they had taken 5 mg DHE three hours before. Four weeks later, sildenafil dose was increased to 100 mg. Following a washout period of four weeks, the crossover type of treatment was started. The clinical observation study was terminated by the administration of 5 mg DHE given again by a routine prescription under the aspect of a non-active drug in the field of sexual activity.

Identity of drugs was known to the participants of the study since all drugs used in this investigation were obtained by routine prescription and commercially available. All information such as effects, sides-effects, bioavailability inherent to the drugs officially used were consequently accessible to the patients.

The valid method of assigning and examining non-responder patients analyzing the study design of the PILRS investigation was to accept some bias such as no drug blinding, one examiner, open reporting system.

After the run in period, patients were assigned to their treatment groups by forming two blocks of therapy A or B, one started on Monday or Thursday with 50 mg sildenafil (group A), the second block starting on Tuesday or Friday with 50 mg sildenafil after 5 mg DHE had been taken three hours before (group B). The washout period is an integral part of the crossover procedure (group A therapy was switched to group B therapy and *vice versa*). The control group received 5 mg DHE as a non-active drug in the field of sexual activity.

The valid selection of the dose regimen was recommended for sildenafil and DHE as routinely used. Most of the patients had already experience in the range of 50–10 mg of sildenafil, so the dose increase from 50 mg to 100 mg after four weeks was not new to patients, since in case of non-responsiveness the drug elevation has been recommended in the officially available drug information.

Drugs allowed before and during the study were necessary for the treatment of concomitant diseases, except for drugs containing nitrate or nitro formulation used for the therapy of heart diseases. Most of the drugs used as concomitant therapy had already been administered during the first use of sildenafil while experiencing non-responsiveness.

The specific efficacy variables of the PILRS investigation were measured according to the International Index of Erectile Function (IIEF). This questionnaire is currently in use worldwide and has been published in Goldstein et al. [14].

The adverse event data analysis is obtained by questionnaire during the medical exam and by rating after being discussed with the patients.

Categorizing and scoring of all efficacy and safety events were regarded as accurate by the study management. Primary efficacy variables according to PILRS investigation are the rating profile of IIEF 3, 4, 5, and 6 with application of the international guidelines, the critical efficacy – in the current investigation the missing therapeutic effect, or the no change expectancy. The study management verified the results by applying a second rating system, that of Wellness scores. No change in primary and secondary efficacy variables would therefore signify the functional accuracy of the evaluation systems based on the implementation of informational quality control system.

Statistical and analytical basis of the PILRS investigation: As described by the study management, the statistical analysis would prove no difference in the observed group under the assumption that groups treated with monotherapy or in combination show no difference. This test procedure was applied to designate $p < 0.01$ as significant. As the analysis was primarily related to the severity scores of IIEF, the clinical results representing these scores were also accepted. The statistical analysis between the different observation groups was primarily intended to record. The first hypothesis was that there is no difference between the inter-assay groups studied. According to the study design the intra group analysis was also performed, because each group received the same drug prescription in a cross over procedure after wash out period. By this second hypothesis no significance in the intra-assay analysis would be expected. Further to statistical threshold values obtained from the second recorded system, the Wellness rating system was thought to further back up the inter – and intra group calculation.

Outcome of the PILRS study with regard to PDE5 inhibitor combination therapy in patients with ED

Analyzing the study results of the PILRS investigation showed that 77 out of the 81 patients qualified for the study in both therapy groups A and B with a defined randomization pattern and a baseline IIEF profile of IIEF3: 1,1 ± 0.2,

IIEF4: 1.2 ± 0.4, IIEF5: 1.2 ± 0.4 IIEF6: 1.7 ± 0.4, the adverse events in both groups were n = 17 and n = 18 respectively, causing no protocol violation.

The data set analyzed from the PILRS investigation exhibit a clear difference in the treatment groups with sildenafil and DHE compared to baseline values to IIEF 3: 3.0 ± 0.7 IIEF 4: 2.9 ± 0.6 IIEF 5: 2.8 ± 0.6 IIEF 6: 3.2 ± 0.5 for 50 mg sildenafil/5 mg DHE and IIEF 3: 3.8 ± 0.7, IIEF 4 $3.5 \pm 0.8 \pm 0.6$ IIEF 5: $3,6 \pm 0.7$ IIEF 6 $4.0 \pm 0,8$ for 100 mg sildenafil/5 mg DHE, respectively.

To discuss the information collected in detail it is of interest that patients with low to non-responsiveness show a slight increase in their mean value, especially in the therapy group with 100 mg sildenafil. This however supports the hypothesis of no change tested when statistical analysis is applied. All analysis contemplated in the primary values of the IIEF 3, 4, 5, 6 are statistical without significance and support the finding of no effect between groups treated with 50 or 100 mg sildenafil. The associated confidence interval displays a similar range to the values during the run in period. There is a clear overlap to prove the zero hypothesis of no change. In the case of the additionally prescribed compound – DHE– added to sildenafil the zero hypothesis of no change is no longer valid. These values are quite distinct; as shown by the statistical analysis, there is a significant change in patients (N = 47, therapy procedure A, starting, (Mon-Thu group) treated with sildenafil 50 or 100 mg in combination with DHE. It is statistically significant when the group under combination therapy is compared in its IIEF Profile 3, 4, 5, 6 to the pretreatment values or even to the values under 50 or 100 mg sildenafil alone. Without sildenafil, the DHE has simply no effect on the ED. The percentage of values given for patients receiving the therapy 3 is valid for the number of patients that display better values compared to the total number of the group A. In group B corresponding results were obtained.

In the PILRS study, a second parameter is introduced to countercheck the IIEF results, pretreatments results and no-to-low responding results ranged not higher than 2.4 ± 0.7 while improvement by the effective therapy displayed values as high as 3.9 ± 1.1.

All patients analyzed in their corresponding therapy group do show only for the combination therapy a significant increase of their scores. The results of wellness categories match to the presented IIEF categories and their results.

The statistical/analytical issues of the PILRS investigation apply the two-sample t-test with the two sample sizes of equivalence with different standard deviation. By this method, results of two groups were compared to demonstrate the null hypothesis of significant difference. The significance level (i.e. the p-value) was expected in the range of $p > 0.05$ that would prove the group categorical profile values in similar range.

On the contrary, p-values <0.01 will demonstrate the positive response to treatment.

Groups treated with sildenafil (50, 100 mg) and DHE (5 mg) in combination showed a significant improved profile of all IIEF 3–6 parameters when compared to groups treated with sildenafil or DHE alone.

High risk reduction by PDE5 inhibitor combination therapy in non-responder patients

The study management demonstrated the analysis of the safety evaluation of the PILRS investigation with no severe side effects above the expectancy profile due to treatment with sildenafil alone (50–100 mg). More accurately, the current data displayed a lower categorical factor scoring of adverse events compared to the use in the first order of the drugs as a single agent.

According to ICH Guideline on Clinical Safety Data Management, there were no serious adverse events in any of the patients observed under each drug alone, or in combination. Similarly, no significant adverse effects were reported so far inducing withdrawal from the prescribed drugs.

Summarizing the adverse events by categorical scoring with consecutive statistical data analysis, a significant decrease of these events in patients under combined treatment (sildenafil and DHE) have been reported (p-values < 0.001) when taking headache, visus disturbances, gastralgia, and diarrhea into account.

The statistical implication of the safety evaluation shows that the combination of sildenafil and DHE improves the safety profile of the drug administered for erectile dysfunction.

Conclusion for non-responders from the PILRS study

According to the study management efficacy and safety results of the clinical observation with sildenafil (50–100 mg) alone, or in combination with 5 mg DHE proved a new effective combination possessing two important benefits:

- reintroduction of response to sildenafil in non-responders to monotherapy, and
- decrease of adverse events by frequency and degree compared to the treatment with sildenafil alone.

Although these findings were unexpected from a clinical standpoint the PILRS is of great importance for the status of integrated erection (IES). The clinical relevance in the light of severe adverse events in patients with heart disease taking PDE5 inhibitors with concomitant nitrate is even more dangerous if the restrictive dosage rules for PDE5 inhibitors are followed. Only combination with DHE would therefore, as indicated by this study, reduce the risk of inadvertently or intentionally ingested PDE5 inhibitors with nitrate compounds. It appears that the dual drug targeting is changing primarily the circulation as well, as there are secondary changes in the urethra, prostate and pelvic region, restoring the integrated erection status (IES).

Clinical rationale for a PDE5 inhibitor combination therapy in patients with ED

Based upon the increase of local NO by sexual stimulation, PDE5 inhibitor induces high cGMP levels with consecutive muscular relaxation and increase of the blood influx rate in the penile corpora cavernosa. A defective venous drainage, however, will increase the efflux from the corpora cavernosa and the blood pooling level cannot be retained, demonstrated by artificial erection during a cavernosogram in elderly patients. Mechanical obstruction of the penile venous system significantly prolongs erection, a device that is utilized by vacuum pump with a valve.

If the small venous vascular muscular ring of the peripenile area is reacting similarly to PDE5 inhibitor by NO and cGMP increase, induction of venous relaxation will drain blood from the corpora cavernosa. The venous circulation in the pelvis is further deteriorated under PDE5 inhibitors.

Patients under direct injection of adrenergic drugs into the corpora cavernosa have a significant increase of the pressure of the sphincter externus urethrae, as well as of the sphincter internus urethrae losing their erection; a procedure used in patients during surgery of transurethral resection who have a prolonged erection. Direct injection of adrenergic drugs will reduce the erection within 15–25 min and simultaneously increase the pressure of the sphincter internus and externus of the urethra. Consequently, the flow from peripenile veins will be decreased. The local efflux during cavernosography is reduced; the pressure profile in the bladder neck region increases almost to 200% under sympathicomimetic, to support an integrated erection (IES), when both drugs, the PDE5 inhibitor as well as the adrenergic compound are given orally [13].

Supportive evidence from experimental work for a PDE5 inhibitor combination therapy in patients with ED

As proven by experimental research the content of NO in dilatated veins such as the spermatic veins in variocele testis are increased. Similarly the concentration of alpha adrenergic receptors in varicosis seems to be elevated. Consequently there will be a negative effect of phosphodiesterase subtype 5 Inhibitors on the peripenile venous systems with increased draining of blood by the veins of the pelvis [15].

Venous constriction will be expected in pathological cavernosograms and reduced venous drainage by the pelvic veins will follow therapy of sympathico-adrenergic compounds. NO levels in the internal spermatic vein were 36.05 ± 8.92 micromol/l, compared to 19.41 ± 4.12 micromol/l in the peripheral vein and the difference was statistically significant ($p < 0.01$). The authors conclude in view of the results, increased NO levels in the dilated varicocele vein might be responsible for spermatozoa dysfunction [16].

Single versus dual approach of ED in respect to PDE5 inhibitor combination therapy in patients with ED

Monovalent drug approach cannot completely induce the multi-cascade process of sexual activity in multimorbid men with ED [13] as shown by the PILRS study.

This clinical problem is immanent for all drugs from the first generation of PDE5 inhibitors (i.e., sildenafil), the second-generation PDE5 inhibitors (i.e., vardenafil and tadalafil), dopamine receptor agonists (i.e., apomorphine), and drugs for both intracavernous and non-intracavernous administration (e.g., prostaglandin E1 [PGE1; alprostadil]).

Although first-line treatment with PDE5 inhibitors is well established, playing a central role in regulating smooth muscle tone, and controlling a wide variety of physiologic processes, but there is no effect on the regional venous constriction by these drugs. The PDE5 inhibitors increase the physiologic response to sexual stimulation by modulating and affecting the cellular responses required for erectile smooth muscle relaxation. Specifically, PDE5 inhibitors inactivate PDEs that normally metabolise the nucleotide cyclic GMP (cGMP). The accumulation of cGMP allows enhanced smooth muscle relaxation and improved erection through increased blood flow, but this is counterproductive to cAMP level in the regionally coordinated veins to the corpora cavernosa.

Although the studies by Halcox [17] and Mahmud [18] identified sildenafil as dilatatory agent with improved constriction response to acetylcholine when endothelial dysfunction was present in the coronary arteries, we would advocate using sympathicomimetic drugs in combination with PDE5 inhibitors to reduce all types of side effects as shown by the PILRS study.

Our studies show that sildenafil plus DHE also improves sexual function in men with chronic health conditions who are at increased risk of developing resistance to monovalent ED-therapy. In addition, sildenafil has shown some efficacy in treating ED in men with diabetes [4], depression [19], and spinal cord injury [20], conditions that might be further improved by sildenafil plus DHE in case of low response to therapy.

In terms of safety, data indicate that sildenafil is well tolerated [21]. The most common side effects include facial flushing, headache, nasal congestion, dyspepsia, and dizziness [21, 22]. Vision problems, including color differentiation and blurriness, have also been reported [5, 21].

Summary of the future for PDE5 inhibitor combination therapy in patients with ED

The complex mechanism of local, regional and central regulation for erectile processing consists of an integrated biochip interaction procedure with some identified processors such as NO, PDE5, penile smooth muscle mass, VIP, rhoA- Kinase, D2 or more specific D4 receptor, other unidentified components and biochemical software.

Subnormal abnormalities in genetic make up [22], and as well disease-induced dysfunction will affect erectile processing at any stage possible and at all levels susceptible creating a complicated clinical feature of non-responders and responders to restore the integrated erection status (IES).

From our observation, it can be inferred that the drug combination will cover the same clinical indication with lower side-effects as seen in the PILRS study. The dual drug approach will also reduce the risk of cardiac side-effects, including sudden cardiac death, ventricular arrhythmia, cerebrovascular hemorrhage, transient ischemic attack, myocardial infarction, and hypertension mainly in multiple drug abusers.

Especially in countries with low medical standards, the abuse of PDE5 inhibitors in chronic health condition may be high asking for a safer primary therapy of ED. Conditions including recent history of myocardial infarction, cerebrovascular accident, life-threatening arrhythmia, significantly elevated blood pressure above 170/100 mm Hg or hypotension, a history of unstable angina or cardiac failure, may be disregarded [5, 21]. Other recent labelling instructions also recommend that men who are taking alpha-blockers to treat their high blood pressure or prostate problems should not take sildenafil in a dose greater than 25 mg within four hours of taking the alpha-blocker. A drug which combines PDE5 inhibitor with a sympathicometic compound will, according to the PILRS study, reduce this risk. However, cardiac risk is not thought to be a problem for most people taking sildenafil as previously discussed, and the American College of Cardiology and the American Heart Association have reported a joint consensus statement to this effect [23]. This statement however is not true for drug abusers (multiple drugs inclusive pop-up drugs) who may suffer from sudden heart problems after sildenafil.

The second generation of PDE5 inhibitors includes clinically vardenafil and tadalafil, which provide more versatility in rate of onset and activity duration when compared with sildenafil [24].

Comparing sildenafil Onset (OP) period: 30–60 min, Duration period (DP) 4 h to that of Vardenafil (OP) 25–30 min (DP) 4–5 h or that of Tadalafil (OP) 16–60 min DP < 36 hours [25–27] may ask to detailed clinical studies when using the dual drug approach to treat ED successfully for integrated erection (IES).

As expected tardenafil also has similar side-effects, the most common include headache, dyspepsia, back pain, nasopharyngitis, and nasal congestion similar to other PDE5 inhibitors. Analysis of the incidence of MI across all studies shows that of more than 4000 tadalafil-treated patients (including those in double-blind and open-label studies), six patients had a myocardial infarction, five of whom were enrolled in the open-label studies [25–27].

Tardanfil is also contraindicated in patients with nitrate therapy, so that a combination with DHE is expected to improve the safety profile of tardenafil and finally of all PDE5 inhibitors. The exploitation of dopamine receptor agonists such as DHE in the dual drug approach is strongly recommended in all patients suffering from ED to achieve partially or completely integrated erection IES.

References

1 Braun M, Wassmer G, Klotz T, Reifenrath B, Mathers M, Engelmann U (2000) Epidemiology of erectile dysfunction: results of the 'Cologne Male Survey'. *Int J Impot Res* 12: 305–311
2 Meuleman EJ (2002) Prevalence of erectile dysfunction: need for treatment? *Int J Impot Res* 14 (Suppl 1): S22–S28
3 Johannes CB, Araujo AB, Feldman HA, Derby CA, Kleinman KP, McKinlay JB (2000) Incidence of erectile dysfunction in men 40 to 69 years old: longitudinal results from the Massachusetts male aging study. *J Urol* 163: 460–463
4 Dey J, Shepherd MD (2002) Evaluation and treatment of erectile dysfunction in men with diabetes mellitus. *Mayo Clin Proc* 77: 276–282
5 Carbone DJ, Jr., Seftel AD (2002) Erectile dysfunction: diagnosis and treatment in older men. *Geriatrics* 57: 18–24
6 Carson CC (2002) Erectile dysfunction in the 21st century: whom we can treat, whom we cannot treat and patient education. *Int J Impot Res* 14 (Suppl 1): S29–S34
7 Shabsigh R (2000) Economic aspects of erectile dysfunction. In: H Jardin, G Wagner, S Khoury, F Giuliano, H Padma-Nathan, R Rosen (eds): *Erectile Dysfunction*. Health Publication Ltd, Plymouth, UK, 55–66
8 Brock G, Carson CC, Giuliano F et al. (2002) Symposium: A New Choice for First Line Therapy. Program and abstracts of the 5th Congress of the European Society for Sexual and Impotence Research; December 1–4, 2002; Hamburg, Germany
9 Shabsigh R, Klein LT, Seidman S, Kaplan SA, Lehrhoff BJ, Ritter JS (1998) High incidence of depressive symptoms is associated with erectile dysfunction. *Urology*: 52: 848–852
10 Stief CG (2002) Is there a common pathophysiology of erectile dysfunction and how does this relate to new pharmacotherapies? *Int J Impot Res* 14 (Suppl 1): S11–S16
11 Lee IC, Surridge D, Morales A, Heaton JP (2000) The prevalence and influence of significant psychiatric abnormalities in men undergoing comprehensive management of organic ED. *Int J Impot Res* 12: 47–51
12 Araujo AB, Johannes CB, Feldman HA, Derby CA, McKinlay JB (2000) Relation between psychosocial risk factors and incidence ED: prospective results from the Massachusetts Male Aging Study. *Am J Epidemiol* 152: 533–541
13 Dunzendorfer U, Behm A, Dunzendorfer E, Dunzendorfer A (2002) Drug combination in the therapy of low response to phosphodiesterase 5 Inhibitor in patients with erectile dysfunction. *In Vivo* 16: 345–348
14 Goldstein I, Lue TF, Padman-Nathan H, Rosen RC, Steers WD, Wicker PA (1998) Oral Sildenafil in the treatment of rectile dysfunction. *N Engl J Med* 338: 1397–1404
15 Ozbek PJ, Turkoz Y, Godgeniez R, Ozourlu F (2000) Increased nitric oxid production in the spermatic vein of patients with varicocele. *Eur Urol* 37: 72–75
16 Miller VM, Rud KS, Gloviczki P (2000) Pharmacological assessment of adrenergic receptors in human varicose veins. *Int Angiol* 19: 176–183
17 Halcox JP, Nour KR, Zalos G, Mincemoyer RA, Waclawiw M, Rivera CE, Willie G, Ellahham S, Quyyumi AA (2002) The effect of sildenafil on human vascular function, platelet activation, and myocardial ischemia. *J Am Coll Cardiol* 40: 1232–1240
18 Mahmud A, Hennessy M, Feely J (2001) Effect of sildenafil on blood pressure and arterial wave reflection in treated hypertensive men. *J Hum Hypertens* 15: 707–713
19 Katz SD, Balidemaj K, Homma S, Wu H, Wang J, Maybaum S (2000) Acute type 5 phosphodiesterase inhibition with sildenafil enhances flow-mediated vasodilation in patients with chronic heart failure. *J Am Coll Cardiol* 36: 845–851
20 Stuckey BG, Jadzinsky MN, Murphy LJ, Montorsi F, Kadioglu A, Fraige F, Manzano P, Deerochanawong C (2003) Sildenafil citrate for treatment of erectile dysfunction in men with type 1 diabetes: results of a randomized controlled trial. *Diabetes Care* 26: 279–284

21 Nurnberg HG, Seidman SN, Gelenberg AJ, Fava M, Rosen R, Shabsigh R (2002) Depression, antidepression therapies, and erectile dysfunction: clinical trials of sildenafil citrate (Viagra) in treated and untreated patients with depression. *Urology* 60 (2 Suppl 2): 58–66
22 Sperling H, Eisenhardt A, Virchow S, Hauck E, Lenk S, Porst H, Stief C, Wetterauer U, Rübben H, Müller N, Sifferts (2003) Sildenafil response is influenced by the G protein ß3 subunit GNB3 C825T Polymorphism. *J Urol* 169: 1048–1051
23 Derry F, Hultling C, Seftel AD, Sipski ML (2002) Efficacy and safety of sildenafil citrate (Viagra) in men with erectile dysfunction and spinal cord injury: a review. *Urology* 60 (2 Suppl 2): 49–57
24 Padma-nathan H, Eardley I, Kloner RA, Laties AM, Montorsi F (2002) A 4-year update on the safety of sildenafil citrate (Viagra). *Urology* 60 (Suppl 2B): 67–90
25 ACC/AHA Expert Consensus Document (1999) Use of sildenafil (Viagra) in patients with cardiovascular disease. *Circulation* 99: 168–177
26 Brock GB, McMahon CG, Chen KK, Costigan T, Shen W, Watkins V, Anglin G, Whitaker S (2002) Efficacy and safety of tadalafil for the treatment of erectile dysfunction: results of integrated analyses. *J Urol* 168 (4 Pt 1): 1332–1336
27 Thadani U, Smith W, Nash S, Bittar N, Glasser S, Narayan P, Stein RA, Larkin S, Mazzu A, Tota R et al. (2002) The effect of vardenafil, a potent and highly selective phosphodiesterase-5 inhibitor for the treatment of erectile dysfunction, on the cardiovascular response to exercise in patients with coronary artery disease. *J Am Coll Cardiol* 40: 2006–2012

Extended clinical use of sildenafil in patients with IPP, prostatitis and infertility syndrome

Mirko Müller

Department of Urology, University of Düsseldorf, Moorenstrasse 5, 40225 Düsseldorf, Germany

Sildenafil citrate (Sc) use in patients with Peyronie's disease

Peyronie's disease (PD) is a localized connective tissue disorder, characterized by changes in the collagen composition of the tunica albuginea of the penis. As a consequence alterations of the tunica albuginea may dramatically affect erectile function. Erectile dysfunction (ED) is known to occur in 20–40% of men with PD because of penile fibrosis, resulting in decreased arterial inflow and veno-occlusive dysfunction. ED is not always a result of changes in penile structure alone, but painful erections and psychological elements associated with increased penile curvature such as performance anxiety may also contribute to this fact. Furthermore, the disease occurs in middle-to-late age, where ED is a common problem in general [1–5].

Despite the relative increase in the incidence of PD since sildenafil citrate (Sc) was introduced in 1998, there is a lack of studies regarding sildenafil treatment in patients with PD and associated ED. A medline research using the keywords "Sildenafil" and "Peyronie's disease" revealed only one article addressing this topic [6]. Maybe the producers statement of caution in the package insert 'Viagra [SC] should be used with caution in patients with anatomical deformation of the penis (such as angulation, cavernosal fibrosis or Peyronie's disease)' led many to believe that this topic is already settled.

Thus, a study by Levine et al. is to this author's knowledge the only one that has evaluated sildenafil efficacy in PD patients. The authors evaluated safety and efficacy of SC in patients with ED associated with PD using a questionnaire. In the evaluation 73 patients of the study population were given a prescription for SC. Of these patients, 75.0% returned the questionnaire while 5.5% did not fill their prescription and 6.8% did not engage in sexual activity following an initial trial of SC due to side effects (flushing, headaches).

According to the questionnaire responses, 34 of 48 (70.8%) patients reported that they were either very satisfied or somewhat satisfied, five (10.4%) patients were neither satisfied nor dissatisfied, and nine (18.8%) patients were somewhat dissatisfied or very dissatisfied with the effectiveness of SC in enhancing their erectile response. No patient reported a worsening of penile deformity or an increase in penile pain. Levine concluded that there appears to be no contraindication using SC as representing the least invasive and most convenient treatment option for ED in PD. Although the potential risk of coital trauma to the erect penis with PD is present, there is no evidence from this study that erections and coitus enhanced specifically by SC resulted in worsened deformity or progression of the PD. In summary questionnaire results reveal that SC is an agent that allowed successful coitus in 70.8% of males with PD [6].

Sildenafil citrate use in patients with prostatitis

Prostatitis is truly a major healthcare problem. Up to 50% of men at some point in their life may be affected by it [7]. Prostatitis affects adult men of all ages. Data from the National Center for Health Statistics demonstrated that in the US there are actually more physician visits for prostatitis than for benign prostatic hyperplasia or prostate cancer [8].

The incidence and prevalence of prostatitis are believed to range from 5–8% [9–12]. At the scientific level, the causes of prostatitis are incompletely understood. Controversy remains among the most highly regarded experts. Patients may have more than one cause operating at the same time, or prostatitis could be several different diseases that present with the same or similar sets of symptoms in different individuals.

Acute prostatitis is well known and nearly always easy to treat. Erectile dysfunction (ED) is not a significant problem in such cases. In contrast, ED is much more prevalent/relevant in chronic prostatitis. A workshop on chronic prostatitis sponsored by the National Institutes of Health/National Institute of Diabetes, Digestive and Kidney Diseases (NIH/NIDDK) in 1995 created a new classification system that provides standard definitions for the different types of prostatitis (Tab. 1 [13]).

The new definition recognized that pain was the main symptom in abacterial chronic prostatitis, and is therefore the optimal criterion to differentiate chronic prostatitis patients from control patients or patients experiencing other genitourinary problems.

Nonbacterial prostatitis respectively chronic pelvic pain syndrome (CPPS) is eight times more common than acute and chronic bacterial prostatitis [8, 14]. Organisms such as *Chlamydia trachomatis*, *Ureaplasma urealyticum*, *Mycoplasma hominis* and *Trichomonas vaginalis* have been postulated as etiologic agents. CPPS typically presents with pain, urinary urgency, weak stream, nocturia, urinary hesitancy and sexual dysfunction (e.g., painful ejaculation, post-ejaculatory pain and hematospermia). The condition is easily confused with benign prostatic hyperplasia (BPH). Other than in the case of BPH, physical examination of the prostate does not reveal abnormal parameters. In a recent study, Beutel et al. found a higher prevalence of sexual dysfunction, especially erectile dysfunction (42%), libido loss (36%), and premature ejaculation (26%) in patients with CPPS [15]. Mehik et al. found ED in

Table 1. The NIH prostatitis classification system

Category	Designation	Status of infection
I	Acute bacterial prostatitis	Acute infection of the prostate
II	Chronic bacterial prostatitis	Recurrent infection of the prostate
III	Chronic non-bacterial prostatitis/ chronic pelvic pain syndrome	No demonstrable infection
IIIA	Inflammatory	White blood cells in semen/EPS/post-prostatic massage urine
IIIB	Non-inflammatory	No white blood cells in semen/EPS/post-prostatic massage urine
IV	Asymptomatic inflammatory prostatitis	No subjective symptoms detected either by prostate biopsy or the presence of white blood cells in EPS during evaluation for other disorders

42% and loss of libido in 24%. Psychological distress is frequent in these patients and quality of life is often impaired [16].

To the author's knowledge there is no published study with respect to the use of sildenafil in patients with chronic prostatitis, respectively with CPPS. Despite the worldwide problem of non-bacterial prostatitis and associated erectile dysfunction, there is a lack in systematic evaluation of the treatment of ED in this disease with sildenafil. In the absence of such studies, sildenafil remains the first-line treatment in patients with erectile dysfunction.

Sildenafil citrate use in male infertility

Infertility is the inability to conceive after at least one year of unprotected intercourse. Since most couples are able to conceive within this time, physicians recommend that those unable to do so be assessed for fertility problems. In men, hormone disorders, illness, reproductive anatomy trauma and obstruction, genetic and epigenetic defects, as well as sexual dysfunction can temporarily or permanently affect sperm and prevent conception.

In addition, involuntary childlessness is considered to be a chronic stressor for couples suffering from infertility. Stress itself may interfere with spermatogenesis and decrease the fertility rate. Long periods of diagnostic and treatment procedures may also have a negative impact on the sex life of the infertile couple. In fact, males in such circumstances suffer more frequently from sexual disturbances taking the form of erectile dysfunction, ejaculatory disorders, loss of libido and a decrease in the frequency of intercourse. The use of sildenafil can help to partially overcome these sexual symptoms [17–19].

Sildenfafil citrate has no effect on both macroscopic and microscopic seminal parameters [20–22].

References

1 Hellstrom WJ (2003) History, epidemiology, and clinical presentation of Peyronie's disease. *Int J Impot Res* 15 (Suppl 5): S91–2. Review
2 Wunderlich H, Werner W, Schubert J (1998) Incidence of induratio penis plastica and erectile dysfunction. *Urol Int* 60: 97–100
3 Sommer F, Schwarzer U, Wassmer G, Bloch W, Braun M, Klotz T, Engelmann U (2002) Epidemiology of Peyronie's disease. *Int J Impot Res* 14: 379–383
4 Jarow JP, Lowe FC (1997) Penile trauma: an etiologic factor in Peyronie's disease and erectile dysfunction. *J Urol* 158: 1388–1390
5 Pryor P, Ralph DJ (2002) Clinical presentations of Peyronie's disease. *Int J Impot Res* 14: 414–417
6 Levine LA, Latchamsetty KC (2002) Treatment of erectile dysfunction in patients with Peyronie's disease using sildenafil citrate. *Int J Impot Res* 14: 478–482
7 Stamey T (1980) Urinary tract infections in males. In: Stamey T (ed): *Pathogenesis and treatment of urinary tract infections*. Williams and Wilkins, Baltimore, 342–429
8 Roberts R, Lieber M, Bostwick D, Jacobsen S (1997) A review of clinical and pathological prostatitis syndromes. *Urology* 49: 809–821
9 Moon TD (1997) Questionnaire survey of urologists and primary care physicians diagnostic and treatment practices for prostatitis. *Urology* 50: 543
10 Moon TD, Hagen L, Heisey DM (1997) Urinary symptometiology in younger men. *Urology* 50: 700
11 Roberts RO, Jacobsen SJ, Rhodes T, Girman CJ, Guess HA, Lieber MM (1997) A community based study on the prevalence of prostatitis. *J Urol* 157: 242A
12 Egan KJ, Krieger JL (1997) Chronic abacterial prostatitis a urological chronic pain syndrome? *Pain* 69: 213
13 National Institutes of Health (1995) Summary statement. National Institutes of Health/National Institute of Diabetes and Digestive and Kidney Diseases workshop on chronic prostatitis. Bethesda, MD
14 Badalyan RR, Fanarjyan SV, Aghajanyan JG (2003) Chlamydial and ureaplasmal infections in patients with nonbacterial chronic prostatitis. *Andrologia* 35: 263–265
15 Beutel ME, Weidner W, Brähler E (2004) Chronic pelvic pain syndrome and comorbidity, *Urologe A* 43: 261–267. German
16 Mehik A, Hellström P, Sarpola A, Lukkarinen O, Järvelin MR (2001) Fears, sexual disturbances and personality features in men with Prostatitis: a population-based cross-sectional study in Finland. *BJU Int* 88: 35–38
17 Lenzi A, Lombardo F, Salacone P, Gandini L, Jannini EA (2003) Stress, sexual dysfunctions, and male infertility. *J Endocrinol Invest* 26 (3 Suppl): 72–76
18 Tannini EA, Lombardo F, Salacone P, Gandini L, Lenzi A (2004) Treatment of sexual dysfunctions secondary to male infertility with sildenafil citrate. *Fertil Steril* 81: 705–707
19 Kaplan B, Ben-Rafael Z, Peled Y, Bar-Hava I, Bar J, Orvieto R (1999) Oral sildenafil may reverse secondary ejaculatory dysfunction during infertility treatment. *Fertil Steril* 72: 1144–1145
20 Aversa A, Mazzilli F, Rossi T, Delfino M, Isidori AM, Fabbri A. (2000) Effects of sildenafil (Viagra) administration on seminal parameters and post-ejaculatory refractory time in normal males. *Hum Reprod* 15: 131–134
21 Purvis K, Muirhead GJ, Harness JA (2002) The effects of sildenafil on human sperm function in healthy volunteers. *Br J Clin Pharmacol* 53 (Suppl 1): 53S–60S
22 Du Plessis SS, De Jongh PS, Franken DR (2004) Effect of acute *in vivo* sildenafil citrate and *in vitro* 8-bromo-cGMP treatments on semen parameters and sperm function. *Fertil Steril* 81: 1026–1033

The cultural impact of sildenafil

Jan Dunzendorfer[1], Udo Dunzendorfer[2], Harald Förster[2]

[1] Humboldt University Berlin, Kulturwissenschaftliches Institut, Sophienstr. 22a, 10178 Berlin, Germany
[2] J.-W.-G.-Universität, Zeil 65–69, 60313 Frankfurt/Main, Germany

Sildenafil, the New Age drug

Mammalian sexuality consists of two main components, that responsible for continuation of the species (reproduction) and that of spontaneous desire (libido). Reproduction is vital to maintain the species, although without a libido component there would be insufficient stimulus. Therefore, biological reproduction can only function with adequate libido, which has the sole biological function of initiating and promoting sexual reproduction.

Mammalian sexual drive is dominated by hormones. In many cases, the employment of sexual drive in mammals is linked to a repeatedly periodic cycle, namely the rut. When not in rut, then the sex drive is suppressed. Humans do not "rut" – their sex drive is always present! Compared to and in contrast to animals, the human being is aware of the consequences of sexual desire (libido) and can therefore control it. Thus in the Prologue to Faust, Goethe lets Mephistopheles argue (internet):

- Earth's little god retains his same old stamp and ways
- And is as singular as on the first of days.
- A little better would he live, poor wight,
- Had you not given him that gleam of heavenly light.
- He calls it Reason, only to pollute
- Its use by being more brutal than any brute.

Goethe's Mephistopheles considered and reduced reason (a gift of the gods) to the ability to cause anxiety and consternation of the libido-orientated human, namely man. In this sense the human being does not require reason to control earthly desire, but considering and realising the possible consequences, he would be forced to more or less repress it. Therefore, reason is a Danaer gift in the eyes of Mephistopheles. In Faust, Gretchen and humanity perish as a result of reason not controlling but promoting man's libido.

As opposed to female mammals, males require a hormone-mediated body alteration prior to copulation, namely the erection of the phallus. The erect phallus is not only in primitive races seen as a symbol of manhood (phallus cult); male virility (not fertility) is seen as a graven image. As a rule, the development of higher civilisation is combined with a decline or even with active combating of the phallus cult, sometimes to complete repression or ban. Of all cultures, Christianity has perhaps gone the furthest: even in modern times, naked sexuality would become a vice; sex drive and the bodily contact between man and woman should have only one aim, namely reproduction. The New Testament does not mention sexuality – it has only the one message that man, built in the shape of God, is spiritual – God is both spirit and reason. Augustine considered that recognition of nudity and therefore shame was the result of the first earthly human sin followed by expulsion from Paradise (the Garden of Eden). The spirit – according to Christian belief – or the divine reason is standing against the animal libido. Reason and libido are opposites and thus not compatible, struggling against one another. Humans, in particular man, can with reasoning decide whether he controls libido and the body or whether libido and the body dominates the man. It's not reason or thought which differentiates man from animals, rather the implementation!

Man has used reason to explain the biological processes which are linked with reproduction. Thus, the use of such reasoning has led to the development of perfect contraception, allowing complete separation of sexual desire (libido) from the sole idea of reproduction. Through scientific knowledge, reproduction has been recognized as the unwanted side-product of sexual drive, the result of which is the consequent circumvention of reproduction. Sexual desire (libido), separated from the idea of reproduction, is an end to itself. Therefore the declining birth rate is to be seen as the end of human existence, the triumph of body and sex over mind and reason, thus leading to the decline of mankind. This trend to complete separation of sexual desire from the idea of procreation was envisaged by Aldous Huxley in the 1930s in his book *Brave New World*. The future of a lust-orientated race which has lost spiritual control consists of artificial wombs and the planned and mechanised creation of humans in test-tubes! Mother and father are no longer necessary. The man is reduced to a sperm donor, the woman to an egg donor. Babies no longer see their first light at birth, rather by the uncorking of a glass bottle! In this nice new world, pregnancy is a sacrilege or a crime. Is this Utopia, where sexual freedom is not enslaved by pregnancy, a dream, wishful thinking or a nightmare?

The more or less accidental finding that sildenafil has an erectile effect has contributed to the accelerated development of a modern "phallus cult". Sildenafil, as a vasodilator, may well be aimed at controlling the phallic erection; however, the question is whether a diminished erectile capacity can be regarded as a disease in the clinical sense. Basically it must be assumed that for the affected person, the capacity to maintain an erection is not vital at least. The lack of erectile capacity (impotence) can, in certain men, lead to psychological effects and *vice versa*. Goethe dedicated a poem to his obvious knowledge of impotence: You bad guy, how motionless you are, cheating your master the highest fortune. It is significant that Goethe regards his virility (i.e. the erectile capability of the phallus) as "the greatest fortune of Mankind"!

Sometimes impotence is related to an underlying disease such as diabetes or to stress. In such cases it may seem medicinally correct to treat impotence so as provide a reproductive capacity. Basically though, such "aphrodisiacs" are used exclusively to increase the frequency of sexual urge. These preparations are considered as additives (life style medications) to support what is seen as a modern way of life.

In *Symposium*, Plato with the boy Phaedros emphasised the importance of Eros for man with men (!) or had he purposely overemphasised the point. "It is to decide if the higher fortune for the guy is to be a good lover for the darling or the better to be a good darling for the lover."

On the other hand, in "The Republic", Plato envisaged the female as the "reproductive carrier" following the practice in Sparta of allowing only qualified men to reproduce. Would Plato in one of his Symposia have recommended sildenafil to increase man's sexual desire? Would it have been a real advice or just ironic? Socrates, married to Xanthippe, would possibly have relied on the effects of sildenafil. Epikur emphasised a quiet, well-tempered life free of stress and his idea of desire did not include strenuous sex, therefore sildenafil would not have been necessary for this way of life. The great physician Hippocrates, devisor of the Hippocratic Oath, would certainly have not prescribed sildenafil. The ancient Greek society at the time of the Philosophers was relatively liberal and the Dionysus cult was a modified Phallus cult. Alcohol would relieve sexual inhibitions and was thought to increase potence. In Imperial Rome, the stoic Seneca, Nero's tutor, was a voice in the wilderness calling to his citizens: "They should definitely stop trying to bring together what cannot be brought together, namely Virtue and Desire. It's a fault, reinforcing the prejudices of that corrupt contemporary individuals."

Virgil found the better note in his prose. The historian Suetonius wrote moral stories about the figures of the immoral Roman Emperors. In ancient Rome as well as in ancient Greece, sildenafil would have been the last thing needed to accelerate the decline of these ancient cultures.

In the Old Testament, God is occasionally seen to chastise immoral mankind. For example, He sent the Flood and He destroyed Sodom and Gomorrah. On the other hand, He appeared not to intervene in the certainly immoral actions of King David or of King Salomon. Orthodox Judaism has clear principles: "Only he that has a wife can study the Tora with a pure mind. For the sake of prostitution, every man has his wife, says the Prophet. Early marriage is an important factor in preventing fornication."

Sex yes, but only in wedlock! The New Testament sees the salvation of mankind in extreme self-discipline and in strict denial of worldly desires. Christianity is the sole religion which propagates the absolute supremacy of mind over body. According to christian-humanistic belief, human reason exalts mankind above the senseless animals and therefore reason can be seen as an impression of the linking of mankind to God. Even in the latter half of the 20th century, the Catholic Church, in the Enzyklika humanae vitae, condemned contraception. Islam, which is built on the Bible, primarily condemned those societies with phallus cults. Regarding sexuality, the Koran is similar to Judaism. The Koran explicitly regards marriage as the way to save sexually-threatened men. Islamic society is ruled by males and by their interests. That women should be veiled protects man from himself (!) but is only good when all women are veiled! Having many wives was seen as a means of supporting war-widows and orphans. When, according to the Koran, the man is in seventh heaven and has a number of young girls to attend to his demands, the earthly harem is not considered to be degenerative. It is Islam, providing an answer to uninhibited sexuality, which is now feeling its roots threatened by the decadence and exhibitionist phallus cult of modern Western society. Again the struggle of Islam is against the obvious phallus cult, now represented by Western life style and identical with decadent Christianity. It must be assumed that of the three major monotheistic religions only orthodox Christianity (i.e. Catholicism) would condemn and ban the use of sildenafil.

The development of culture and civilisation is an expression of intellect but also that of a partial and at least temporal repression of earthly desire. It can be debated if a society with a phallus cult will develop a civilisation. On the other hand, the development and increasing dominance of a phallus cult was always a sign of the downfall of civilisation and culture. Thought and reason lead to discipline, control of personal desire and control of the animal in man. Mastery of sexual thoughts lifts man above the level of animals. The use of reason exclusively for an uninhibited sexual drive makes man, in the words of Goethe, "more brutal than the brute".

The decline of culture in our affluent society in the 21st century is expressed in an overestimation of sexuality and in a return to a modified and modernized phallus cult. Sildenafil in the modern society is hardly used to treat impotence as a symptom of disease; it is used almost exclusively to increase sexual capacity. Therefore, sildenafil is an expression and a symbol of modern masculinity. Christopher Street Day and the Love Parade are examples of sexual freedom and of an exhibitionistic phallus cult. Sexuality is shamelessly proffered as a "show" and in the existential struggle propagated by Augustine, spirit and reason have lost out to earthly desire and libido. Receivers of social security demand sildenafil (and condoms) for an "appropriate" lifestyle; sexuality and promiscuity is being financed by the community for healthy individuals as the sole sense of life! Huxley's utopian *Brave New World* has become a brutal reality within a short time of less than 100 years. Mankind has sunk from a level of self-control of the sexual drive to the level of animals where the sex drive is governed by the rut. Sildenafil is a drug of changing times, the "New-Age" drug, a symbol of a super-phallus and thus a symbol of a time beyond that of man.

Upgrade in civilization

Global aspects in civilizations inhere a high degree of semantic confusion, which will persist with the unchanged longevity of languages presumably for all futures to come. New acquisitions of cultural, medico-technical or biopharmacological assets will primarily not improve the understanding of the individual nor of the society being exposed when importing foreign ideas or products [1, 2].

When however new assets as a drug are directly intruding to everybody's mind powered by medial transmission, high sales rate will immediately be commented by all persons involved, discussing the medical, personal and civilisation consequences to the local society.

Friendly interest in new foreign ideas or products would not always be welcomed in former times and civilization; there might have been tight restriction particularly to drugs interfering with the sexual sphere in humans.

A sphere with extensive impact on social life equally sensitive for all promising procedures [4–6].

In communities with a high degree of collectivism and non-individuality, such as in the Middle Ages, in Islamic societies, in China during the time of exclusion, in modern fundamentalist sovereignties, a restrictive control of drugs changing the behaviour of the sexual sphere can be predicted. A limited access presumably for the exclusive leader in that civilization may be the results of stringent political control of drugs.

In may therefore be doubted that in those communities with powerful religious controls a commercial success of PDE5 inhibitors used as drugs for changing the behaviour of the sexual sphere would be shown [7–9].

It makes no sense to analyse a situation of free access of particularly drugs interfering with the sexual sphere in humans such as PDE5 inhibitors in any society, but if commercially available in Athens at the time of Plato and Aristotle, the drugs would have been extensively studied and analysed and might have changed some vital attitude in Greek understanding of men in various areas of private, cultural, social, and political life.

In contrast to the Athens polis the drug certainly would not be discussed in Sparta nor in any society at war or in a post-war situation, as seen in German for instance during 1947. At that time sociocultural regions of interest were covered by existential needs in a chaotic economic and family situation [10].

There are other historical periods the drugs might have readily distributed, during rococo or during the time before the French Revolution, while being used in the aristocratic society the increased cost of living might have accelerated the uproar and the beginning of the French Revolution, evoked by reports on the extreme luxury of life at the Sansouscis [11].

Christianity asserts the priority of metaphysical love over individualized love, a theorem that can even nowadays only with difficulties be accepted. Drugs such as PDE5 inhibitors used for changing the behaviour of the sexual sphere stabilizing men or in some case even effective frigid women sexual desire must create problems. As individual love is a constant part of the metaphysical love with a high priority in theological and ethical dimension in the life of human, a drug personalizing the behaviour of the sexual sphere presents a constant threat in contemporary Christianity, because of its effects to the sacrosanct borders of men. Therefore Christianity will induce tremendous efforts to elucidate the autonomised love of men and try to reintegrate and or to transform it to its former metaphysical properties [12].

Islamic teaching encourages Muslims to live their life in any country and under different cultures having always in mind the Islamic umma. In a more strict understanding a drug changing the behaviour of the sexual sphere by reinforcing the power of men is enthusiastically accepted and the drug will not have any problems as long as men will stay with Islamic women, and the promotion of the family and the communal wellbeing are achieved. But there is a major drawback, that this drug presents a Western world product, which consequently will have severe toxic implication to the Islamic umma [13].

Jewish community life and ethic are based on long tradition with the concrete element, that the only way towards God is unitarily inherited to them. The central vision of life even when deprived as often in their history, presenting frequently as a history of diaspora, no external philosophers or extrinsic psychoanalysis are recognised as complementary elements. A non-Jewish resident will not be regarded as equal partner unless there is a change to Judaism. Drugs elements changing the behaviour of the sexual sphere will not possess sufficient power to overcome the strict belief and power of the kabbala and in addition these drugs bearing foreign components as non-kosher residues or nanoparticle of that are not admitted by the Torah [14–15].

The Chinese perception of the world order was stable as a Sinocentric topos for centuries, Western invasion by car, drugs and high technology has completely changed society, but the understanding, and evaluation by the government or by the Chinese population has not been issued. It would be a progress if drugs changing the behaviour of the sexual sphere can and will replace ancient mysterious rules and drugs elements made from rhinos and other animal which have nothing in common with humans [16].

Love interfering medication is best launched in a society with classical liberalism as is emerged in modern states. Here individualism and personal insight have come to a certain degree of elaboration. Following the impact of religious obedience, obligation, strict regulation of community life, the individual will see for rescue measurements. Dependant on the local structure the fees for classical liberalism will be diplomatic, convincing by arguments on elaborated personal feelings and observations. Officially confronted with those elaborated problems a society with classical liberalism will respond dependant on its facilities. Referral systems for personal exit of problems i.e., by means of psychoanalysis, philosophy, religion is a quality pattern of classic liberalism. The subjective crash of individuals quite differently handled in societies such as Japanese, Hindu, African or Latin American has its greatest chance in an internal exit of problems from previous traditional positions. Contrary to traditional ethics liberal thinking cherishes a position which abandones ancient features. If necessary, this active procedure may be supported by drugs to trespass the borders of regulation. Within the optimal decision others of the community are not endangered. Individuals on the road want a morally secure zone free from incessant predation by other partner in their society. Initiated by Renaissance, further propagated in the Enlightment of the Classic period and nowadays improved by network reference classic liberalism is always under pressure by fanatism and intolerance as well by low intellectual standards and under mediocre to low budget economic situations. Medication seems to have an important role in such tight social perturbances, but can medication also improve problems of intimacy in the civilized and/or in the natural world of men [17–23]?

Further Reading

Augustinus Aurelius (1978) *Vom Gottesstaat* (deutsch von Wilhelm Thimme). Artemis, Zürich/München, 2. Auflage.
Huxley A (1991) Schöne neue Welt. 58. Auflage. Insel Verlag, Frankfurt
Platon (1991): *Symposium und Politeia*. Übersetzt von F. Schleiermacher, Herausgeber: K. Hülser. Insel Verlag, Frankfurt
Preuss J (1911) Biblisch-talmudische Medizin. Fourier, Wiesbaden
Seneca (1996) Von der Seelenruhe. Herausgeber: Heinz Berthold, Bechtermünz Verlag, Augsburg

References

1. Merck & Co (1899) Mercks's Manual. Merck & Co, New York, 137–138
2. Lanier ML (1896) L'Asie. Berlin Frères, Paris
3. Porst H (1999) Die gekaufte Potenz. Steinkopff Verlag, Darmstadt
4. Blech J (2001) Wann ihr wollt. In: Stiftung Deutsches-Hygiene Museum (Hrsg) (Red. H Raulff): Sex – Vom Wissen und Wünschen. Ruit, Ostfildern, 187–195
5. Sigusch V (1998) Das simple Prinzip von Ursache und Wirkung funktioniert nicht. In: Wissenschaft und Technik. Frankfurter Rundschau 134: 134–137
6. Zizek S (1999) Wer naiv fragt, wird schockiert. Die Zeit 50: In Themen der Zeit
7. Roth G (2004) Kant und Hirnforschung. Forschung und Lehre 3: 132–133
8. Huntington S (2002) The clash of civilization. The Free Press, London
9. Seibt F (1987) Glanz und Elend des Mittelalters. Siedler Verlag, Berlin
10. Der Spiegel (1947) Der Spiegel 10, 13, 14,19: 15, 17, 19
11. Weiss P (2002) Die Verfolgung und Ermordung Jean Paul Marats, dargestellt durch die Schauspielgruppe des Hospiz. Edition Suhrkamp, Frankfurt
12. Miller RB (2001) Christian attitude towards boundaries. In: D Miller, SH Hashmi (eds): Boundaries and Justice. Princeton University Press, Princeton, 15–37
13. Zaman MR (2001) Islamic perspectives on territorial boundaries and autonomy. In: D Miller, SH Hashmi (eds): Boundaries and Justice. Princeton University Press, Princeton, 182–202
14. Novak D (2001) Land and People. In: D Miller, SH Hashmi (eds): Boundaries and Justice. Princeton University Press, Princeton, 213–236
15. Grözinger KE (2003) Kafka und die Kabbala. Philon Verlags GmbH, Berlin
16. Nylan M (2001) Boundaries of the body and body politics in early confucian thought. In: D Miller, SH Hashmi (eds): Boundaries and Justice. Princeton University Press, Princeton, 111–135
17. Steiner H (2001) Hard Borders, Compensation, and Classical Liberalism. In: D Miller, SH Hashmi (eds): Boundaries and Justice. Princeton University Press, Princeton, 79–88
18. Grassi E (1957) Kunst und Mythos. Rowohlt, Hamburg

19 Kubie LS (1956) Psychoanalyse ohne Geheimnis. Rowohlt, Hamburg
20 Bahrdt HP (1961) Die moderne Großstadt. Rowohlt, Hamburg
21 Tadiè JY, Tadiè M (2003) Im Gedächtnispalast. 2Cotta'sche Buchhandlung, Stuttgart
22 Duerr HP (1990) Intimität. Suhrkamp, Frankfurt
23 Morris DB (2000) Krankheit und Kultur. Kunstmann GmbH Verlag, München

Index

abnormal vision 42
achalasia 18
adrenergic drug 177
adverse event 42, 76, 176
adverse events, decrease of 176
age-associated changes in male
  sexual response 103
ageing male 90
alcohol use 108
allosteric site, PDE5 23
alprostadil 53
amantadine 111, 112
amlodipine 157
amygdala 68
ancient mysterious rules, chinese
  perception 192
antidepressant 102, 108, 111
antidote 111
antihypertensive medication 35,
  157, 158
aphrodisiacs 189
apomorphine 77, 106, 107
apoptosis 52
arousal 117, 118, 120–124
arrhythmia 153, 156
arrhythmia, PDE5 inhibitors 156
ASIA scores 72
audio-visual sexual stimulation
  (AVSS) 73
Augustine 188
autonomic dysreflexia 76

Barrington's nucleus 68
BDI 110
benign prostatic hyperplasia (BPH)
  44, 109, 110
biochemical potency 23
bladder dysfunction 74

α-blocker 41, 147
β-blocker 41
bulbocavernous reflex (BCR) 72, 73
bupropion 111
buspirone 111, 112

cAMP 15, 163
cardioprotective effects of PDE5
  inhibitors 159
cardiovascular disease 108, 129,
  135, 151–159
cardiovascular disease, patient
  counceling 154
cardiovascular disease, treatment of
  ED with PDE5 inhibitors 155,
  156
cardiovascular disease and
  contraindications, PDE5 inhibitor
  157
cardiovascular drugs and PDE5
  inhibitors 155–158
cardiovascular drugs and sildenafil
  155, 156
cardiovascular effect 26
cardiovascular mortality and PDE5
  inhibitors 158, 159
cardiovascular risk and sexual
  activity 152–154
cardiovascular safety 135
catalytic domain, PDE5 21
catecholamines 102
Caverject 107
cavernous nerve 51, 57
cavernous nerve interposition
  grafting 57
cavernous nerve reconstruction 57
central nervous system 68
CES-D 109

cGMP 15, 21, 36, 54, 103, 106, 129, 156, 163
childlessness 185
Chinese perception 192
christian-humanistic belief 188, 189
chronic dialysis 91
chronic heart failure 131
chronic obstructive lung disease 166
chronic pelvic pain syndrome (CPPS) 184
chronic stable coronary artery disease 152
chronic treatment with PDE5 inhibitors 159
chronicle device 153
cimetidine 106
classical liberalism 192
clinical interview 110
clomipramine (CMI) 105, 106
CMI therapy 106
Cologne Male Survey 170
combination therapy, treatment of non-responders 172
co-morbidity, cardiovascular diseases 151, 152
co-morbidity, models of 108
congestive heart failure 44, 152, 153
contraception 188
coronary artery dilation 41
coronary artery disease 133, 153, 159
coronary heart disease 152, 155
coronary heart disease, PDE5 inhibitors 155
cross-talk 19
culture, decline 190
cyclic adenosine monophosphate (cAMP) 15, 163
cyclic guanosine monophosphate (cGMP) 15, 21, 36, 54, 103, 106, 129, 156, 163
cyclic GMP-binding site 21
cyclic GMP-specific phosphodiesterase type 5 (PDE5) 106

CYP2C9 158
CYP3A4 158
cyproheptadine 111
cytochrome P450 system 158

defective regional venous flow 170
depression 39, 70, 71, 75, 92, 108–112
depression and sexual dysfunction in men 108–112
DHE 175, 180
DHE, dual drug approach 180
diabetes mellitus 72, 90, 134
diabetic patient 38
diaspora 192
Dionysus cult 189
dizziness 42
dopamine 68, 102
dopaminergic drugs 102
dose titration 42
drug development 5–7
drug interaction 143
dyspepsia 42
dysphoria 109
dysreflexia 76

endothelial dysfunction 131
Enzyklika humanae vitae 190
Epidemiologic Studies Depression Scale (CES-D) 109
epidemiology 170
Epikur 189
erectile capacity 188
ED, psychosocial implications 84
ED etiology, mixed 38
ED etiology, organic 38
ED etiology, psychogenic 38, 39
erythromycin 106
estrogen 119, 122–124
excretion 148
exercise treadmill testing (ETT) 152

failure rate, elderly men 171
fanatism 193
female sexual dysfunction 44

French Revolution 191
frenulum squeeze procedure 105

Ginkgo biloba 111
global efficacy assessment 37
glutamate 68
Goethe 187
granisetron 111
Greek understanding of men 191
guanyl cyclase 19
guanylate cyclase 163
guilt feelings 85

Hamilton Rating Scale for Depression (HAM-D) 110
headache 42
Health Professionals Follow-up Study 151
heart failure 44, 131, 152, 153, 155, 159
heart failure, PDE5 inhibitors 155
hermatospermia 184
higher civilization 188
hippocampus 68
Hippocrates 189
HIV protease inhibitor 157, 158
Huxley 188
hypertension 38, 157–159
hypertensive patient 157
hypoactive sexual desire disorder (HSDD) 103–105
hypogastric nerve (HGN) 68, 69
hypogonadism 108
hypotension 156, 157
hypothalamus 68
hypoxia 164

ICH Guideline on Clinical Safety 176
identity of drugs 173
iloprost 159
immunophilin 59
Imperial Rome 189
infertility, male 185
inflammatory bowel disease 93

integrated erection status (IES) 172
International Index of Erectile Function (IIEF) 50, 72
interstitial lung disease 165, 166
intracavernosal injection therapy 54, 107
ischemic heart disease 42

Johns Hopkins Sexual Behaviors Consulting Unit 109
Judaism 189, 190

kidney transplant recipient 91
King David 189
King Salomon 189

libido 184, 187
libido loss 184
locus coeruleus 68
lower motoneuron lesion (LMN) 69
lower urinary tract symptoms 44
lumbosacral motor nuclei 105, 111

male erectile dysfunction (MED) 15, 17, 20
mammalian sexuality 187
marriage 190
Massachusetts Male Aging Study (MMAS) 109, 151, 152, 170
maximum serum concentration 43
major depressive disorder (MDD) 108, 109, 111, 112
measurement of male erectile response 108
medical risk factor 171
medication, discontinuation 42
medulla oblongata 68
menopause 117, 120, 121, 123, 124
Mephistopheles 187
mesolimbic system 71
midbrain 68
mirtazepine 111
monoamine oxidase inhibitor (MAOI) 107, 111

monovalent drug approach 178
mood 109, 112
Multinational Survey of the Aging Male (MSAM-7) 170
multiple regulatory mechanism 171
multiple sclerosis (MS) 39, 70, 72, 76
multiple system atrophy (MSA) 91
myocardial infarction (MI) 42, 152
myosin light chain kinase (MLCK) 103
myosin light chain phosphatase (MLCP) 103

nebivolol 147
nefazodone 111
nerve sparing 50
nerve sparing radical prostatectomy 50
nervi erigentes 69
nervous system 102
neuroplasticity 69
neuroprotection 58
neurotrophic factors 59
neurovascular bundle 51
neurovascular physiology of erectile function 103
new targets, treatment of ED 172
New Testament 188
nicorandil 147
nitrate 106, 144, 156, 157
nitric oxide (NO) 19, 20, 36, 54, 102, 103, 106, 111, 129, 156, 157, 163, 177
NO-donator 106, 156, 157
NO-synthase 36, 163
nitroglycerin 26
nitrovasodilator 20, 26
nocturnal penile tumescence (NPT) 73, 108
nocturnal penile tumescence and rigidity testing (NPTR) 73
non-kosher residues 192
non-responder 170, 172
norepinephrine (NE) 68, 102

nucleus paragigantocellularis 68

Old Testament 189
onset after ingestion, sildenafil 43
open non-blinded study 173
organic nitrate 156, 157
oxytocin 68

PAH 163, 164
parapyramidal reticular formation 68
paraventricular nucleus 68
Parkinson's disease (PD) 70, 72, 76, 91
pathological cavernosogram 177
patient counceling regarding sexual activity and cardiovascular disease 154
pause-squeeze technique 105
penile constriction device 75
penile corpus cavernosum 26
penile deformity 107
penile fibrosis 183
penile intracavernosal injection 110
penile prosthesis 77, 107
penile self-injection therapy 107
penile tumescence 108
penile vascular smooth muscle 20
penile vasculature 20
penile vibrator stimulation 75
pervasive anhedonia 111
Peyronie's disease 107, 183
phallus cult 188
pharmacokinetic properties, sildenafil 27
pharmacotesting 73
PDE5 inhibitor, abuse of 179
PDE5 inhibitors and cardiovascular diseases 154–159
PDE6 family 25
PKG 21
platelet aggregation 27
pons 68
post-ejaculatory pain 184
postoperative ED, treatment 53
post-war situation 191

premature ejaculation (PE) 104–106, 184
priapism 107
prostaglandin 103, 171
prostaglandin E1 77
prostate cancer therapy 92
prostatectomy 39, 49, 50, 93
prostatitis, chronic 184
prostatitis, definitions for the different types 184
prostatitis, nonbacterial 184
psychobiology, sexual function 102
psychogenic erection 69, 73–75
psychosexual therapy 75
psychosocial distress 108
psychosocial health 95
psychostimulant 111
psychotropic drugs 94, 105, 110–112
pudendal nerve 72
pudendal somatosensory evoked potential 74
pulmonary arterial hypertension (PAH) 18, 44, 135, 163, 164

QRS complex 41
QT dynamics 156
quality of life (QoL) 43, 83, 85, 87, 95
QoL, assessment instrument for erectile function 87
quality of partnership (QoP) 83, 85, 87, 95
QoP, assessment instrument for erectile function 87

radical prostatectomy 40, 49, 93
raphe magnus 68
raphe pallidus 68
Raynaud's syndrome 18, 159
rectal cancer, treatment 93
reflexogenic erection 68, 69, 73–75
retrograde ejaculation 68
rhinitis 42
ritonavir 158

rococo 191
Roman Emperors 189

sacrosanct borders of men 192
selective serotonin reuptake inhibitor 39
self-esteem, loss of 95
self-injection technique 107
Seneca 189
serotonergic medication 111
serotonin 39, 68, 102, 111
serum half life 43
sex steroid 102
sexual activity with heart disease 152–154
sexual dysfunction 69, 70, 103, 109, 110
sexual dysfunction, classification 103
sexual dysfunction, impact on mood 109, 110
sexual encounter profile 37
sexual fantasy 104
sexual function 101, 102, 108, 109, 117, 120–123
sexual function, age 101, 102
sexual function, depression 108, 109
sexual function, monoamines 102
sexual function, phases of 101
sexual function, psychobiology 102
sexual history 72
sexual response cycle 101, 102
sexual satisfaction 83
sildenafil, cost-effectiveness 96, 97
sildenafil, cost-utility 97
sildenafil, efficacy 36
sildenafil, public acceptance 96
sildenafil, public opinion 96
sildenafil, safety 36
sildenafil citrate (Viagra) 1–5, 7, 8, 17, 18, 20, 25, 35, 36, 117, 122
sildenafil treatment, negative effects 94
Sinocentric topos 192
smoking 39

social perturbances 193
Socrates 189
Sodom and Gomorrah 189
specific efficacy variables 174
spermatic vein 177
spermatozoa dysfunction 177
sphincter externus urethrae 177
sphincter internus urethrae 177
sphincteric dysfunction 70
spinal cord injury 39, 69, 76, 90
spousal satisfaction 40
SSRI 105, 111, 112
SSRI-associated sexual dysfunction 112
SSRI treatment 111
stable coronary artery disease 153
stroke 18
structured clinical interview 110
Suctonius 189
sural nerve grafting 57
sympathetic nervous system 68, 69
sympathetic skin response (SSR) 74
sympathicomimetic compound 172
systemic blood pressure 27

tachyphylaxis 43
tadalafil 54, 77, 106, 137, 143, 159
testosterone 104, 105, 120, 123, 124
tricyclic antidepressants (TCAs) 111
two-sample t-test 175
Type I diabetes 38
Type II diabetes 159

upper motorneuron lesion (UMN) 69
uremia 91
urodynamics 74

vacuum constriction device (VCD) therapy 107
valvular disease 153, 154
vardenafil 17, 54, 77, 106, 107, 133, 159
varicocele vein 177

vascular endothelium 130
vascular smooth muscle 19, 129
vasoactive intestinal polypeptide (VIP) 103
vasodilation symptoms 42
veno-occlusive dysfunction 183
Viagra® 1–5, 7, 8, 17, 18, 20, 25, 35, 36, 117, 122
Virgil 189
virility, male 188

war 191
wellbeing 83

yohimbine 111

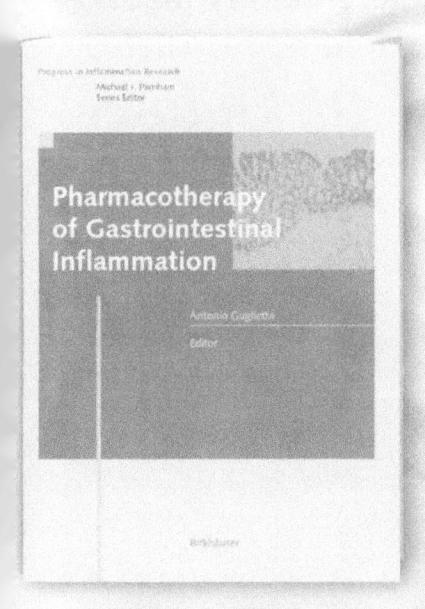

Where quality meets scientific research...

Birkhäuser

Guglietta, A., Grupo Ferrer Internacional, Barcelona, Spain (Ed.)

Pharmacotherapy of Gastrointestinal Inflammation

2003. 156 pages. Hardcover
ISBN 3-7643-6910-8
PIR - Progress in Inflammation Research

In recent years, the area of pharmacotherapy of GI inflammation has witnessed important progresses with new drugs and therapeutic approaches being introduced. The volume reviews the pharmacotherapy of selected gastrointestinal inflammatory conditions chosen on the basis of their clinical importance and the areas where important and exciting progresses have been made recently. Besides discussing current pharmacotherapy to treat the most important GI inflammation conditions, the book also indicates possible future therapeutic avenues likely to become available in a few years.
The book is of interest to various sectors of the scientific community ranging from clinicians to pharmacologists and from biochemists to microbiologists who will find it an useful tool for their clinical practice and research activity.

From the contents:
- *Marco Romano and Antonio Cuomo:* Pharmacotherapy of Helicobacter pylori-associated gastritis
- *Marija Veljaca:* Pharmacotherapy of inflammatory Bowel Disease: novel therapeutic approaches
- *Flavio Lirussi and Beniamino Zalunardo:* Current and future therapy of chronic hepatitis
- *E. Bergogne-Bérézin:* Treatment and prevention of antibiotic associated colitis

For orders originating from all over the world except USA and Canada:
Birkhäuser Verlag AG
c/o Springer GmbH & Co
Haberstrasse 7
D-69126 Heidelberg
Fax: +49 / 6221 / 345 4 229
e-mail: birkhauser@springer.de
http://www.birkhauser.ch

For orders originating in the USA and Canada:
Birkhäuser
333 Meadowland Parkway
USA-Secaucus
NJ 07094-2491
Fax: +1 201 348 4505
e-mail: orders@birkhauser.com

Where quality meets scientific research...

Birkhäuser

Pairet, M. / van Ryn, J., both Boehringer Ingelheim Biberach, Germany (Eds.)

COX-2 Inhibitors

2004. 260 pages. Hardcover
ISBN 3-7643-6901-9
MDT - Milestones in Drug Therapy

The original theory of inducible COX-2 and constitutive COX-1 is continually being updated. COX-2 is not only induced during inflammation but is constitutively expressed in various tissues where it is necessary for physiological functions. It has, however, been conclusively demonstrated that COX-2 inhibitors are important drugs with analgesic and anti-inflammatory activity, and this has confirmed the importance of COX-2 in inflammation.

The discovery of COX-2, its chemical basis and the development of inhibitory drugs are reviewed in this book. Since a large amount of clinical experience with COX-2 inhibitors is now available, the implications of these developments for future therapy are also covered. Additional information on the roles of COX-1 and COX-2 in mucosal protection and ulcer healing also allow a better assessment of the level of improvement achieved by COX-2 inhibitors.

This comprehensive monograph is of interest to clinicians and researchers in academia and industry as well as for advanced students in the field.

From the contents:
- *Regina M. Botting and Jack H. Botting:*
  The discovery of COX-2
- *Lawrence J. Marnett and Amit S. Kalgutkar:*
  Structural diversity of selective COX-2 inhibitors
- *Guenter Trummlitz, Joanne van Ryn and Timothy D. Warner:*
  The molecular and biological basis for COX-2 selectivity
- *K.D. Rainsford:*
  Pharmacology and toxicology of COX-2 inhibitors

For orders originating from the world except USA and C
Birkhäuser Verlag AG
c/o Springer Auslieferungs (SAG)
Customer Service
Haberstrasse 7, D-69126 He
Tel.: +49 / 6221 / 345 0
Fax: +49 / 6221 / 345 42 2
e-mail: orders@birkhauser.cl

For orders originating in the and Canada:
Birkhäuser
333 Meadowland Parkway
USA-Secaucus
NJ 07094-2491
Fax: +1 201 348 4505
e-mail: orders@birkhauser.c

Printed by Libri Plureos GmbH
in Hamburg, Germany